Holger Hoffmann

Abiotic impact of regional climate change on horticultural production

Holger Hoffmann

Abiotic impact of regional climate change on horticultural production

Südwestdeutscher Verlag für Hochschulschriften

Impressum / Imprint

Bibliografische Information der Deutschen Nationalbibliothek: Die Deutsche Nationalbibliothek verzeichnet diese Publikation in der Deutschen Nationalbibliografie; detaillierte bibliografische Daten sind im Internet über http://dnb.d-nb.de abrufbar.
Alle in diesem Buch genannten Marken und Produktnamen unterliegen warenzeichen-, marken- oder patentrechtlichem Schutz bzw. sind Warenzeichen oder eingetragene Warenzeichen der jeweiligen Inhaber. Die Wiedergabe von Marken, Produktnamen, Gebrauchsnamen, Handelsnamen, Warenbezeichnungen u.s.w. in diesem Werk berechtigt auch ohne besondere Kennzeichnung nicht zu der Annahme, dass solche Namen im Sinne der Warenzeichen- und Markenschutzgesetzgebung als frei zu betrachten wären und daher von jedermann benutzt werden dürften.

Bibliographic information published by the Deutsche Nationalbibliothek: The Deutsche Nationalbibliothek lists this publication in the Deutsche Nationalbibliografie; detailed bibliographic data are available in the Internet at http://dnb.d-nb.de.
Any brand names and product names mentioned in this book are subject to trademark, brand or patent protection and are trademarks or registered trademarks of their respective holders. The use of brand names, product names, common names, trade names, product descriptions etc. even without a particular marking in this works is in no way to be construed to mean that such names may be regarded as unrestricted in respect of trademark and brand protection legislation and could thus be used by anyone.

Coverbild / Cover image: www.ingimage.com

Verlag / Publisher:
Südwestdeutscher Verlag für Hochschulschriften
ist ein Imprint der / is a trademark of
OmniScriptum GmbH & Co. KG
Heinrich-Böcking-Str. 6-8, 66121 Saarbrücken, Deutschland / Germany
Email: info@svh-verlag.de

Herstellung: siehe letzte Seite /
Printed at: see last page
ISBN: 978-3-8381-3828-2

Zugl. / Approved by: Hannover, Germany, Leibniz Universität Hannover, Dissertation, 2013

Copyright © 2014 OmniScriptum GmbH & Co. KG
Alle Rechte vorbehalten. / All rights reserved. Saarbrücken 2014

Summary

Climate change will impact horticultural production in the future. Thus, the overarching objective of the present work is to assess the future climatic impact on regional horticultural production by establishing a basic frame of a climate impact modeling chain.

Using high resolved simulated climate time series of future alternatives of the worlds development (SRES emission scenarios B1, A1B, A2), long-term trends of various climate effects on horticultural production were assessed. For this purpose, simulated climate time series were calibrated with observations and effects of resolution, bias, bias correction, scenario, climate model and impact model were investigated. A multidimensional bias correction method was developed in order to optimize climate time series consistency. Furthermore, establishing the simulation chain IPCC-scenario / SRES-emission scenario of greenhouse gases > Global climate projection > Regional climate projection > Bias correction > Climate impact, an ensemble approach consisting of 13 climate projections and 7 phenological models was used to estimate future apple blossom frost risk. Analysis of uncertainty by variance decomposition for climate model, impact model and internal variability in combination with single time series statistics was conducted.

As a result, no increased risk of abiotic factors were found for crop production nor production systems at regional level (Lower Saxony, Germany). However, climate change is likely to lead to a range of changes in horticultural production due to shifts in vegetation period, speed of plant development and growth as well as greenhouse energy demand. Future changes in heat stress and irrigation management are possible. In more detail, future apple blossom frost risk is likely to be at present level or lower, as last spring freeze and bloom will both occur earlier, but with bloom advancing relatively slower than last spring freeze. This effect was attributed to a loss of winter chill, slowing down the advance of bloom due to warming in spring. Hereby the uncertainty of the projection was lowest for temperature, followed by phenology and finally by blossom frost. Although these three target parameters exhibited a minimum of uncertainty in projection for the period 2078-2087, changes in blossom frost risk were lower than internal and model variability. This showed the limits of the meaningful projection horizon for this type of impact study.

Future greenhouse energy consumption was projected, consistently resulting in a mean decrease. Hereby climate was projected to impact mainly beyond mid-century and diverging regionally.

Furthermore, a water stress model was developed and calibrated in order to assess irrigation demand and water stress as exemplified by *Lactuca sativa* var. *capitata* L. While no detailed projection was conducted, the projected mean decrease of summer precipitation cannot be expected to pose a risk to plant production. However, deviations in precipitation patterns should be followed closely.

Finally, methodology as well as risks and trends are reviewed. Specific effects on crop production were found for vegetables with obligate vernalization with delayed vernalization but shorter duration of cultivation with late species of cauliflower.

Keywords: Bias correction, climate change, horticulture

Zusammenfassung

Veränderungen im Klima werden sich auf die gartenbauliche Pflanzenproduktion der Zukunft auswirken. Ziel der vorliegenden Arbeit ist daher, zukünftige regionale Auswirkungen des Klimawandels auf die gartenbauliche Produktion zu simulieren und den dafür notwendigen methodischen Rahmen zu erstellen. Trends verschiedener Klimaeffekte im Pflanzenbau wurden mittels hochaufgelöster simulierter Klimazeitreihen abgeschätzt. Die auf Zukunftsszenarien basierenden Zeitreihen wurden mit Messdaten kalibriert und Effekte von Auflösung, Bias, Biaskorrektur, Szenario, Klima- und Impaktmodell untersucht. Eine mehrdimensionale Methode zur Biaskorrektur wurde entwickelt, um die Konsistenz verschiedener Klimazeitreihen zu optimieren. Basierend auf der Simulationskette IPCC-Szenario / SRES-Emissionsszenario > Globale Klimasimulation > Regionale Klimasimulation > Biaskorrektur > Klimawirkung wurden 13 Klimarealisierungen und 7 phänologischen Modellen verwendet, um die zukünftige Entwicklung des Blütenfrostrisikos bei Apfel abzuschätzen. Unsicherheiten wurden durch Analyse der Varianzen von Klima- und Impaktmodell sowie interner Variabilität als auch durch Statistik einzelner Klimazeitreihen untersucht.
Es wurde kein zunehmendes Risiko für die Produktion auf regionaler Ebene (Niedersachsen) festgestellt. Allerdings wird der Klimawandel wahrscheinlich zu einer Reihe von Änderungen im Gartenbau führen, z.B. durch Veränderung der Vegetationsperiode, Tempo pflanzlicher Entwicklung und pflanzlichen Wachstums sowie Änderungen im Energiebedarf von Gewächshäusern. Änderungen von Hitzestress und Bewässerungsstrategien sind möglich. Im Detail wird das zukünftige Blütenfrostrisiko für Apfel auf gegenwärtigem Niveau oder niedriger liegen, bedingt durch eine langsamere Verfrühung der Blüte im Verhältnis zur Verfrühung des letzten Frühjahrsfrostes. Dies zeigt einen Rückgang der für eine Brechung der Dormanz effektiven Kältestunden, welches die Verfrühung der Apfelblüte bremst. Hierbei zeigte die Projektion der Temperatur die niedrigste Unsicherheit, gefolgt von Phänophasen und zuletzt Blütenfrostrisiko. Diese Größen zeigten ein Minimum an Unsicherheit für den Zeitraum 2078-2087, wobei Änderungen im Blütenfrostrisiko innerhalb der internen sowie Modellvariabilität lagen. Hierdurch wurden die Grenzen dieser Art von Klima-Impakt-Projektion beispielhaft dargestellt. Zudem wurde eine Abnahme im zukünftigen Energiebedarf bundesdeutscher Gewächshäuser projiziert. Eine deutliche wenn auch regional verschiedene Klimawirkung konnte hierbei für die zweite Hälfte des 21. Jahrhunderts festgestellt werden. Ferner wurde ein Trockenstressmodell exemplarisch für *Lactuca sativa* var. *capitata* L. entwickelt und kalibriert, um Änderungen im zukünftigen Bewässerungsbedarf zu ermitteln. Während keine detaillierten Zukunftsprojektionen durchgeführt wurden, stellt eine mittlere Abnahme der Sommerniederschläge vermutlich keine Gefährdung für die gartenbauliche Pflanzenproduktion dar. Dennoch müssen künftige Änderungen im Niederschlagsmuster beachtet werden. Schlussendlich werden Methoden, Risiken und Trends begutachtet. Spezifische Klimaeffekte konnten für den Anbau obligat vernalisierender Pflanzen festgestellt werden, wobei eine verzögerte Vernalisierung sowie eine kürzere Anbaudauer mittlerer bis später Sorten Blumenkohl festgestellt wurde.

Schlagwörter: Biaskorrektur, Klimawandel, Gartenbau

Contents

Summary   1

Zusammenfassung   3

List of figures   9

List of tables   11

List of abbreviations   12

1   Introduction   15
     1.1   Motivation   16
     1.2   Climate impact assessment   18
         1.2.1   General procedure / The IPCC-process   18
         1.2.2   Emission Scenarios   19
         1.2.3   Definition of climate   20
         1.2.4   Climate projection   21
         1.2.5   Ensembles   22
         1.2.6   Bias and bias correction   22
         1.2.7   Aggregation and interpolation   31
         1.2.8   Uncertainties in climate impact projections   33
         1.2.9   Observed and projected climate change   34
     1.3   Susceptibility of plant / horticultural systems to climatic changes   36
         1.3.1   Basic thoughts on climatic impact through changes in distribution parameters   36
         1.3.2   General system parameters   37
         1.3.3   Vulnerable systems   43
         1.3.4   Observed climatic impact   45
         1.3.5   Expected future impact   48

2   General objectives   55

3 Investigations — 57

3.1 Processing and calibration of climate input data — 58
- 3.1.1 Objective — 58
- 3.1.2 Summary — 58
- 3.1.3 Publication: Meteorologically consistent bias correction of climate time series for agricultural models — *Theoretical and applied climatology* — 59

3.2 Future water stress risk for *Lactuca sativa* L. var. *capitata* — 60
- 3.2.1 Objective — 60
- 3.2.2 Summary — 60
- 3.2.3 Publication: Dynamic Modelling of Water Stress for *Lactuca sativa* L. var. *capitata* — *Acta Horticulturae* — 61

3.3 Future bloom and blossom frost risk for *Malus domestica* — 71
- 3.3.1 Objective — 71
- 3.3.2 Summary — 71
- 3.3.3 Publication: Future bloom and blossom frost risk for *Malus domestica* considering climate model and impact model uncertainties — *PLoS ONE* — 72

3.4 Future energy consumption of horticultural production in greenhouses — 91
- 3.4.1 Objective — 91
- 3.4.2 Summary — 91
- 3.4.3 Publication: High Resolved Simulation of Climate Change Impact on Greenhouse Energy Consumption in Germany — *European Journal of Horticultural Science* — 92

4 Closing remarks — 101

4.1 General remarks on presented investigations — 101
4.2 Résumé of specific climate change effects on horticultural production — 101
4.3 Résumé of general climate change effects on horticultural production — 105
- 4.3.1 Projection framework — 105
- 4.3.2 Future trends and risks in horticultural production — 106

4.4 Critical reflexion — 107
4.5 Outlook — 107

5 Bibliography — 109

6 Appendix — 125

6.1 Publications — 126
- 6.1.1 Publications included in the thesis — 126
- 6.1.2 Publications not included in the thesis and conference contributions — 126

6.2 Acknowledgments 128

List of figures

1.1	Climate impact assessment modeling chain	18
1.2	Scheme of emission scenarios	19
1.3	Scheme of a nested model approach	21
1.4	Influence of the choice of horizontal resolution on minimum and maximum area elevation	31
1.5	Basic concept of changes in distribution for normally distributed climate variables	36
1.6	Temperature influence on leaf net photosynthesis and cell cycle	40
3.0	Sketch of developed water stress model for *Lactuca sativa*	65
3.1	Relationship between fresh weight and plant diameter	66
3.2	Stress factor as a function of soil moisture	67
3.3	Irrigation management and development of measured and simulated soil moisture	67
3.4	Influence of the irrigation treatment on measured plant growth and simulated plant growth	68
3.5	Scheme of used input data and projection	76
3.6	Present temperature incidence of Lower Saxony (1991-2010)	79
3.7	Projected changes in air temperature, fulfillment of chilling requirement and onset of flowering	80
3.8	Proportion of years with unfulfilled chilling requirement	80
3.9	Changes in bloom and blossom frost risk as projected by different phenological models and climate runs 1-5	81
3.10	Changes in bloom and blossom frost risk as projected by different phenological models and climate runs 6-13	81
3.11	Changes in last spring freeze	81
3.12	Distribution of projected changes in blossom frost risk by the end of the 21st century (2070-2099 minus 1971-2000) for early and late ripening varieties, phenophases BBCH 60 and 65 and 7 phenological models	82
3.13	Simulated relation between projected absolute changes in decadal mean air temperature and changes in the day of bloom	82
3.14	Uncertainty in the projection of apple bloom (t_2)	82
3.15	Uncertainty pattern of projected temperature (T), apple bloom (t_2) and blossom frost risk (Θ)	83

3.15 Projected changes in greenhouse energy consumption in Germany by 2031–2045 as compared to 2001–2015 for scenarios B1 and A2 and temperature set-points 5/5 °C 95

3.16 Projected changes in greenhouse energy consumption in Germany by 2031–2045 as compared to 2001–2015 for scenarios B1 and A2 and temperature set-points 18/16 °C 96

3.17 Projected greenhouse energy consumption in Lower Saxony simulated for the climate scenario A1B and temperature set-points 5/5 °C (day/night) 96

3.18 Projected greenhouse energy consumption in Lower Saxony simulated for the climate scenario A1B and temperature set-points 18/16 °C (day/night) 97

3.19 Distribution of the projected yearly greenhouse energy consumption in Lower Saxony simulated for day/night temperature set-points of 5/5 °C and 18/16 °C and climate scenario A1B . 97

3.20 Projected yearly greenhouse energy consumption for different day/night temperature set-points, calculated from original and bias corrected simulated climate data in Bremen (scenario A1B) 97

List of tables

1.1	Number of research studies published on climate and climate impact	17
1.2	Emission scenario (SRES) storylines	19
1.3	Climate definitions	20
1.4	Bias correction methods	29
1.5	Examples of sources of uncertainties in climate projections	34
1.6	Global and regional observed and projected climatic changes	35
1.7	Potential "Knock-Out-Effects"	44
1.8	Observed abiotic impact of climate change	46
1.9	Expected trends of future abiotic impact of climate change on plant development or growth	50
1.10	Expected future abiotic impact of climate change on yield or production	53
3.0	Abbreviations (Water Stress)	70
3.1	Published projections of future apple blossom frost risk	74
3.2	Overview of employed data	75
3.3	Phenological models	76
3.4	Denomination of variables and parameters	77
3.5	Stepwise error of simulation chain segments	79
3.6	Prediction Root Mean Squared Error PRMSE of phenological models	79
3.7	Denomination of variables and parameters	89
3.8	Model parameters (early ripeners, BBCH 65, area mean)	90
3.9	Basic simulation input settings for HORTEX	94
3.10	Influence of the bias correction on climate data quality and simulated energy consumption (1977–2010)	97
4.1	Trends and future risks of abiotic impact of climate change for selected horticultural aspects	106

List of abbreviations

The following abbreviations are applied throughout chapter 1 and chapter 4. Abbreviations used in publications included in the manuscript are explained separately if relevant. Symbols not listed are explained where relevant. Regions are abbreviated by standard code (ISO-3166-1 Alpha-2 and ISO 3166-2).

1d		one dimensional
2d		two dimensional
A1		SRES emission scenario, see table 1.2
A1B		SRES emission scenario, see table 1.2
A1F1		SRES emission scenario, see table 1.2 and fig. 1.2
A2		SRES emission scenario, see table 1.2 and fig. 1.2
AR4		Assessment Report No. 4 (IPCC)
AR5		Assessment Report No. 5 (IPCC)
B1		SRES emission scenario, see table 1.2 and fig. 1.2
B2		SRES emission scenario, see table 1.2 and fig. 1.2
bc		bias correction
CAM		crassulacean acid metabolism
cdf		cumulative distribution function
ci		internal concentration of CO_2 in the leaf
CLM		Climate Limited-Area Model (http://www.clm-community.eu/)
CO_2		carbon dioxide
DC		Delta Change approach (see bias correction)
DJF		Winter (December-January-February)
DOY		day of the year, e.g. DOY 41 = February 10
ETa		actual evapotranspiration
ETp		potential evapotranspiration
FACE		Free-air concentration enrichment
GCM		general or global circulation model (e.g. HadCM3, Echam5)
GHG		Greenhouse gas
i		index or time step
IDW		Inverse Distance Weighting
IPCC		International Panel on Climate Change
JJA		Summer (June-July-August)
LAI		leaf area index
LOCI		Local Intensity Scaling (see bias correction methods)
LS		Linear scaling (see bias correction methods)
MAM		Spring (March-May-April)
n		number of elements of a given array

List of abbreviations

NMVOC	non methane volatile organic compounds
NO_x	Mono-nitrogen oxides
n.s.	not significant
O_2	oxygen (molecular form)
O_3	ozone
P	Precipitation
PAR	photosynthetic active radiation
pdf	probability density function
pmf	probability mass function
PT	Power Transformation (see bias correction methods)
$\bar{q}$	mean deviation (bias)
QM	Quantile mapping (distribution based bias correction)
R_G	global radiation
R_{dif}	diffuse radiation
RCM	regional climate model (e.g. REMO, CLM)
RCP	representative concentration pathway
REMO	Regional Climate Model, Max Planck Institute for Meteorology (Hamburg, Germany)
SON	Autumn (September-October-November)
SRES	Special Report on Emission Scenarios, see Nakicenovic et al. (2000)
SWT	soil water tension
T	Air temperature
vpd	vapor pressure deficit
W	World, global
WUE	water use efficiency
x	measured climate variable
$\hat{x}$	simulated climate variable
$\dot{x}$	corrected, calibrated or perturbed climate variable
$X = \{x_1..x_n\}$	observed, measured time series
$\hat{X} = \{\hat{x}_1..\hat{x}_n\}$	simulated time series
$\dot{X} = \{\dot{x}_1..\dot{x}_n\}$	simulated and corrected, calibrated or perturbed time series
$\bar{X}$	mean of observed, measured time series X
$\bar{\hat{X}}$	mean of simulated time series $\hat{X}$
$\bar{\dot{X}}$	mean of simulated and bias corrected time series $\dot{X}$
X_{ref}	observed, measured time series, reference period
$\hat{X}_{ref}$	simulated time series, reference period
$\dot{X}_{ref}$	simulated and bias corrected time series, reference period

$\hat{X}_{fut}$	simulated time series, future
$\dot{X}_{fut}$	simulated and bias corrected time series, future
Δ	Climate signal (Anomaly of climatic variable over time and compared to a reference period, e.g. 1971-2000)
μ_X	location parameter of a given array X (e.g. mean, median)
σ_X	scale parameter of a given array X (e.g. standard deviation, interquartile range)
$\Gamma(k)$	function value of the gamma function

Chapter 1

Introduction

1.1 Motivation

Life develops within its system boundaries. Hence plant development is subject to these boundaries, which can be characterized by external conditions such as air temperature, soil moisture, atmospheric carbon dioxide or radiation, restricting plant development to a specific range of ambient conditions. Moreover these environmental effects exert an influence on the entire system, as optimal development occurs in a narrower range. As a consequence, these circumstances have contributed considerably to the distribution of ecosystems worldwide (Olson et al. 2001). Even though cultivated plants are extracted from their original habitat and have been adapted (e.g. through breeding) to meet horticultural / agricultural demands, growth as well as development and finally yield remain being functions of these circumstances. Hereby climate can be considered as one essentially limiting factor for crop production (Porter and Semenov 2005) and hence for horticultural production, as it determines the required ecological conditions (Krug et al. 2002). Climate, in brief defined as the statistics of atmospheric processes (see tab. 1.3 for further definitions), affects open field production directly, whereas crops in protected cultivation are affected directly only partially, depending on the technical infrastructure[a]. However, in the latter case climate impacts on the technical infrastructure and derived variables, e.g. costs. Any change in climate will therefore lead to changes in the horticultural production, analogously to the observed impact of other affected systems (Hughes 2000). This impact does not necessarily depend linearly on climate change and effects can be enhanced or buffered through either positive feedbacks of compensating mechanisms. Anyhow, increases in global mean air temperature of up to 4.4 °C (Solomon et al. 2007) are expected by the end of the 21st century, compared to the mean of the standard reference period 1971 to 2000. Stating that "temperature affects everything that an organism does" (Clarke 2003) and considering further, that multiple interactions do occur between climate variables and plant responses, it is essential to cope with possible climate impacts on horticultural production in order to satisfy future demands, anticipating possible production risks.

Hereby knowledge on future horticultural production can be gained in several ways, which will be discussed in more detail in the following sections. In brief, this knowledge can be gained through observations combined with simulations. Future atmospheric processes can be simulated to a certain extent physically/dynamically, whereas the climate-plant interface can be assessed with the help of stochastic, mechanistic and empirical models, including further basic considerations based on comparisons with past observations. In order to assess future horticultural production the estimated climate from climate models is subsequently used as input for so-called impact models. This process implies uncertainties in measurement errors, model error concatenation, different resolutions and time scales among others, therefore often requiring a large number of realizations ("runs") in order to depict a more complete picture of possible future scenarios. Hence this computing intensive process (CPU-hours officially used 2012 for the present studies: 222499, RRZN Hannover) has gained increasing attention in the recent past as shown by table 1.1. Increasing funding of research networks in Germany (e.g. KliO (Chmielewski et al. 2009), www.kliff-niedersachsen.de, http://klimzug-nord.de/, http://www.reklim.de/de/, http://www.pa.op.dlr.de/RegioExAKT/), Europe (www.macsur.eu/, http://www.climatechangeintelligence.

baastel.be/piccmat/index.php, http://www.cecilia-eu.org/, http://www.climsave.eu/climsave/index.html) and the world (http://www.agmip.org/about-us/) as well as the IPCC-process (http://www.ipcc.ch/) manifest the rising awareness of this gap of knowledge. An increasing number of climate realizations from regional models of high spatio-temporal resolution have been conducted during the past decade (Solomon et al. 2007), allowing the performance of regional climate impact studies. Despite a wide usage of these climate projections for agricultural purposes, the mentioned knowledge about future horticultural production remains non-satisfying. In the main three gaps arise: 1.) Basic gaps, common to all disciplines, 2.) Unknown effects of transferring methodology from other fields to horticultural science, 3.) Unknown effects specific to horticultural production. Therefore the overarching objective of the present work is to assess the future climatic impact on regional horticultural production by establishing a basic frame of a climate impact modeling chain.

In the following, the present work introduces basic concepts and gives the necessary background for the subsequent research publications. The latter subordinate to the overarching objective, though pursuing their own objectives. It must be added, that although current climatic changes are driven substantially by changes of atmospheric carbon dioxide concentrations, the following work does not include projections of effects of increasing atmospheric carbon dioxide concentration on plant production. Additionally, the work focuses on direct (abiotic) climate effects in order to permit the assessment of simulation uncertainty. Thus climate impact on secondary and more complex effects, such as crop-weed interaction, plant pests or pathogens are not discussed.

Table 1.1: Number of research studies published on climate and climate impact (www.scopus.com, March 2013)

Keywords searched for in 'article title'	Number of studies		
	Period		
	1960-1980	1981-2000	2001-2013
Climate model	86	1,430	4,070
Regional climate model	1	117	783
Climate impact model	1	82	368
Climate + Model uncertainty	1	12	106
Climate impact + Model uncertainty	0	2	28
Climate impact + Agriculture / horticulture	1	22	100
Total number of publications (any field)	3,341,337	11,753,972	17,631,377

[a]Climate of field crops may also be altered through technical measures (e.g. mulch or underground heating)

1.2 Climate impact assessment

1.2.1 General procedure / The IPCC-process

Past and recent climate change can be tracked by means of geology, paleoclimatology as well as meteorology from proxy data and measurements. Accordingly, the corresponding climatic impact can be estimated. On the other hand, future climate is simulated and an impact of future climate change can be assessed through models. An overview on this modeling chain is given in the following.

Large uncertainties about the future development of the driving boundary conditions of the climate have led to the so-called IPCC-process. The Intergovernmental Panel on Climate Change (IPCC) was established in order "to provide the world with a clear scientific view on the current state of knowledge in climate change and its potential environmental and socio-economic impacts" (http://www.ipcc.ch/). Following this process, greenhouse gas emission scenarios were created and used as input in order to drive global circulation models (GCM), (fig. 1.1). Since GCMs operate on a coarse resolution, downscaling is applied by using GCMs output as input for regional climate models (RCM) of higher spatial resolution. The obtained climate time series can be compared to measurements in order to remove systematic errors (bias correction). These time series are subsequently used as input for impact models. Having the climatic impact, risk assessment can be conducted and adaptation strategies can be evaluated.

Additionally the influence of the initial conditions can be estimated by using several climate model runs. Further, ensembles consisting of several GCM-RCM combinations are applied to estimate/reduce the uncertainty of the simulation.

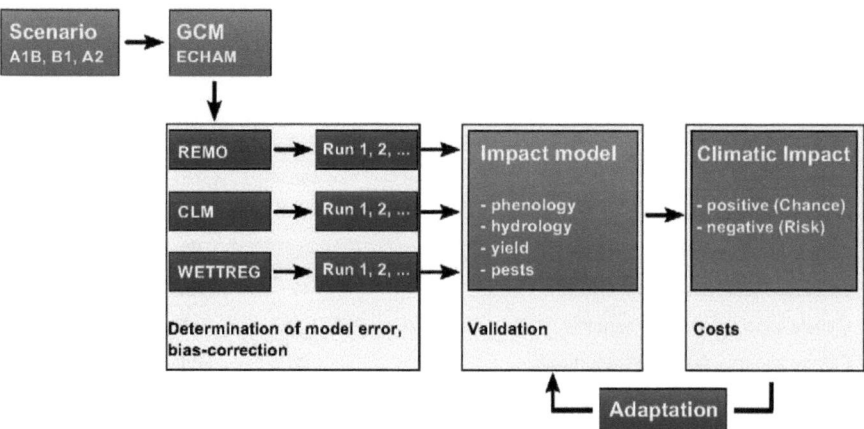

Figure 1.1: **Scheme of a climate impact assessment modeling chain.** Scenarios and models are exemplary.

1.2.2 Emission Scenarios

In order to reach robust decisions, scenarios of the future were developed (Nakicenovic et al. 2000). These scenarios are alternatives of how the future could develop, based on assumptions of demographic and socio-economic development as well as technological innovations. From these assumptions greenhouse gas (GHG) and sulfur emissions are derived and used as input for climate models (radiative forcing) and/or impact models ("CO_2 fertilization", O_3 toxicity). These 40 "SRES"-scenarios are further grouped in 4 qualitative narrative storylines or "families" (A1, A2, B1, B2) and six groups with one illustrative marker scenario each. Hereby the scenarios do not include measures to reduce GHG emissions, but these are reflected in the non-climate change policies of the storylines (Nakicenovic et al. 2000). Furthermore the scenarios have no assigned probability and are supposed to be equally valid. Consequently calculations based on the emission scenarios are denominated projection rather than prediction. The main scenario patterns are given by table 1.2 and fig. 1.2. It must be added, that representative concentration pathways (RCP) (Moss et al. 2010) have been developed and are included in the fifth assessment report of the IPCC (AR5).

Table 1.2: Emission scenario (SRES) storylines

Scenario family	Globalization	Economy	Population	Technological change
A1[a]	homogeneous	rapid growth	peaks approx. 2050, declines thereafter	rapid / more efficient technologies
A2	heterogeneous	regionally oriented	continuously rising	slower than other storylines
B1	homogeneous	rapid change towards service / information economy	peaks approx. 2050, declines thereafter	resource-efficient
B2	heterogeneous	intermediate development	continuously rising (slower than A2)	less rapid, more diverse

[a]Scenario A1B is balanced across energy sources from fossil-intensive to non-fossil, scenario A1F1 is fossil intensive.

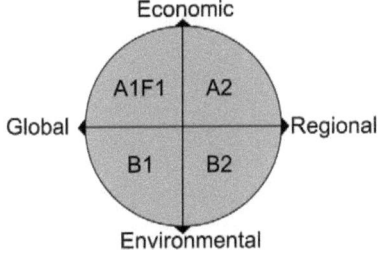

Figure 1.2: Scheme of emission scenarios. Modified from Schroeter et al. (2005)

1.2.3 Definition of climate

Climate has been defined in different ways since 1845 (tab. 1.3). All recent definitions have in common, that climate comprises a long-term and statistical view on atmospheric processes (weather). Hence, for the present work, climate is defined as the statistics of weather. Therefore climate change is the difference in a statistical parameter (e.g. mean, variance) of any period (e.g. 2071-2100) compared to a reference period. The reference period used for the present work is 1971-2000.

Table 1.3: Climate definitions (selection)

Reference	Definition
v. Humboldt (1845: 340)	All atmospheric changes which perceptibly affect our organs[a]
v. Hann (1883)	Entity of all meteorologic phenomena, which characterize the mean state of the atmosphere at a given location of the earths surface.[a,b]
Köppen (1923)	Mean state and usual course of weather conditions at a given location[a,b]
Lorenz (1970)	"collection of all long-term statistical properties of the state of the atmosphere"
Hantel et al. (1987: 1-5)	Statistical behavior of atmosphere, which is characteristic for a relatively large time scale[a]
Solomon et al. (2007: Annex I)	"Climate in a narrow sense is usually defined as the average weather, or more rigorously, as the statistical description in terms of the mean and variability of relevant quantities over a period of time ranging from months to thousands or millions of years. The classical period for averaging these variables is 30 years, as defined by the World Meteorological Organization. The relevant quantities are most often surface variables such as temperature, precipitation and wind. Climate in a wider sense is the state, including a statistical description, of the climate system."
Latif (2009: 13)	Statistical properties of weather[a]

[a] translated from German
[b] as quoted in Bender and Schaller (2012)

1.2.4 Climate projection

Climate is simulated with the help of models, which describe the atmospheric processes over time. While the predictability of weather as a chaotic system is very limited, models may predict changes in the statistics of weather (Latif 2009: 111). Hereby Lorenz (1975) defines predictability of the first kind resulting from the systems initial conditions (e.g. predictability of weather or climate internal variability; Latif 2009: 111) and predictability of the second kind resulting from the boundary conditions (e.g. global climate change estimates; Latif 2009: 111). Hence climate models may predict climate resulting from changes in the climate systems boundary (e.g. radiative forcing).

In order to obtain climate time series of high spatial resolution, a nested approach is used. Hereby global circulation models (GCM) with resolutions lower than 50 km × 50 km are used to drive regional climate models (RCMs) at their boundaries. For instance, fig. 1.3 shows the coarse grid of ECHAM5 (Roeckner et al. 2003) and the nested regional model REMO (Jacob 2001) as well as the area of Lower Saxony (Germany) within the boundaries of REMO. Location and elevation of the latter are depicted as the present work strongly focuses on the regional climate impact of that area.

While REMO is a physical-dynamical climate model, statistical downscaling approaches such as regression methods, weather pattern-based approaches and stochastic weather generators do further exist (Wilby and Wigley 1997). These are based on statistical relationships between large scale and local variables.

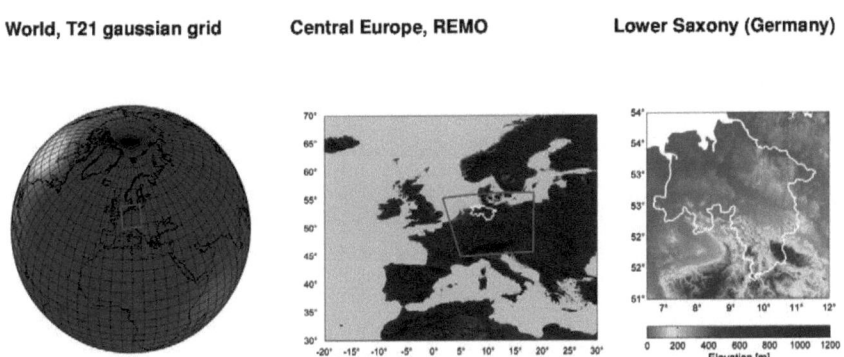

Figure 1.3: Scheme of a nested model approach. A global circulation model GCM, e.g. ECHAM5 (5.6° resolution) drives a regional climate model RCM, e.g. REMO (0.088° resolution), at its boundaries (red frames, left and middle). Subsequently regional climate data from the RCM is used for impact studies, e.g. for Lower Saxony (white frame, middle and right). Elevation and coastlines were obtained from the GLOBE Task Team, Hastings et al. (1999); National Geophysical Data Center (2013).

1.2.5 Ensembles

In order to cope with uncertainties and to minimize the influence of chaotic processes of the climate system on the projected climate, several projections are conducted (realization or run). These ensembles are used to estimate the climate (impact) signal, being the mean change compared to a reference period, as well as to estimate the uncertainty of the projection. In weather forecasts, this approach can be used to estimate the reliable horizon of prediction. If a large number of ensemble members coincides in prediction, the forecasted situation is more reliable. Analogously, in climate projections ensembles are used to estimate the influence of the boundary conditions. For this purpose, runs with different initial conditions are conducted.

As climate models exhibit different errors, model ensembles are further used. Hereby the uncertainty of the different climate models is estimated and noise (internal variability) reduced (Déqué et al. 2007; Ruosteenoja et al. 2007). For instance, four or five models are necessary to estimate precipitation changes (Giorgi and Coppola 2010). Assuming that model errors are random, the model average can be an estimator of the climate signal (Latif 2009: 132). However, according to Hawkins and Sutton (2009) this should not be assumed. Finally, super-ensembles of models and scenarios can be obtained through scaling approaches (Ruosteenoja et al. 2007).

1.2.6 Bias and bias correction

Simulated climate time series deviate from measurements and observations (Hoffmann and Rath 2011). Due to model errors, spatial resolution and interpolation method as well as data processing (e.g. Richter correction of measured precipitation for undercatch; Richter 1995) deviation may occur in the mean as well as in the distribution of the time series. In the following time series are defined as arrays of measured climate variables $X = \{x_1..x_n\}$, as arrays of simulated climate variables $\hat{X} = \{\hat{x}_1..\hat{x}_n\}$ and as arrays of bias corrected, perturbed or calibrated climate variables $\dot{X} = \{\dot{x}_1..\dot{x}_n\}$. Array means are indicated by bars (e.g. $\bar{X}, \bar{\hat{X}}, \bar{\dot{X}}$) whereas time series sections or elements of the reference period (e.g. years 1971-2000) and any period in the future (e.g. years 2071-2100) are indicated by subscripts ref and fut (e.g. X_{ref}) respectively. For denominations see also the list of abbreviations on page 12.

The mean deviation from measured time series and simulated time series is the bias or systematic error:

$$\bar{q} = \frac{\sum_{i=1}^{n} \hat{x}_i - x_i}{n} \quad \text{with} \tag{1.1}$$

$\bar{q}$: bias

x_i : observed climate variable at time step i

$\hat{x}_i$: simulated climate variable at time step i

i : time step

n : number of time steps i

Hereby biases occur on all timescales (e.g. seasonal precipitation bias) and are regarded as underestimation (negative bias) or overestimation (positive bias) of a model. Deviations in the distribution function of a climate variable hence are under- or overestimation of a given range, e.g. overestimation of low precipitation ("drizzle") by climate models.

Biases of climate model time series influence the accuracy of the projected climatic impact, as the latter is projected with the help of impact models calibrated with measured time series. In the case of impact models which depend linearly on a climate variable (e.g. temperature sum based growth model based on monthly mean temperature without thresholds), the error of the projected impact will change proportional to the bias. However, most impact models consist of non-linear equations and are therefore susceptible to distribution based errors. For example, the estimation of frost risk for a given crop requires accurate reproduction of temperatures ≤ 0 °C in addition to accurate mean temperatures. Consequently, small deviations in the distribution of x and $\hat{x}$ may add up to large errors of the projected impact. For instance, in the mentioned example, a climate model may underestimate mean temperatures while also underestimating frost occurrences, hence leading to the erroneous assumption of no frost risk. Therefore climate model biases have been largely studied at different resolution and bias correction approaches have been developed.

Bias correction in its simplest version can be conducted by shifting each value of a simulated time series by the bias itself. This implies corresponding measured/observed climate time series and can be formulated as:

$$\dot{x}_i = \hat{x}_i - \bar{q} \quad \text{with} \tag{1.2}$$

$\dot{x}_i$: bias corrected climate variable at time step i

$\hat{x}_i$: simulated climate variable at time step i

$\bar{q}$: bias

Extendending this concept, the variance can be further included (modified from Ho et al. 2012):

$$\dot{x}_i = \mu_X + \frac{\sigma_X}{\sigma_{\hat{X}}} \cdot (\hat{x}_i - \mu_{\hat{X}}) \quad \text{with} \tag{1.3}$$

μ_X : location parameter of observed climate time series X (e.g. mean)

$\mu_{\hat{X}}$: location parameter of simulated climate time series $\hat{X}$ (e.g. mean)

σ_X : scale parameter of observed climate time series X (e.g. standard deviation)

$\sigma_{\hat{X}}$: scale parameter of simulated climate time series $\hat{X}$ (e.g. standard deviation)

These approaches (eq. 1.2,1.3) are referred to as "bias correction" in a narrower sense by Ho et al. (2012) and Hawkins et al. (2013) as the variability of the produced time series $\dot{X}$ originates in the variability of the simulated time series. In this sense, the perturbation of future time series in order to remove systematic errors by adding a climate signal to observed climate data is named "calibration". However, if no further specified, in the present work all procedures to correct simulated time series by implying statistical moments obtained from measured time series (mean, variance and/or skewness) are referred to as bias correction, following literature in the main, e.g. Piani et al. (2010); Haerter et al. (2011); Hagemann et al. (2011). A review on the main bias correction methods is given by Teutschbein and Seibert (2012).

A correction method for future climate time series is the so-called **delta-change approach** (DC), adding the projected climate change ("delta") to a measured time series. This procedure can be applied either using an absolute (e.g. for temperature; Seaby et al. 2013) or a relative change (e.g. for precipitation or evapotranspiration; Ines and Hansen 2006; Seaby et al. 2013). The application of an absolute change ("delta") can be written as (modified from Seaby et al. 2013):

$$\dot{x}_i = x_i + \Delta \quad \text{with} \tag{1.4}$$

$$\Delta = \hat{\bar{X}}_{fut} - \hat{\bar{X}}_{ref}$$

$\dot{x}_i$: perturbed ("calibrated") climate variable at time step i

x_i : measured climate variable at time step i

Δ : climate signal

$\hat{\bar{X}}_{ref}$: mean of simulated climate variable of reference period

$\hat{\bar{X}}_{fut}$: mean of simulated climate variable of future period

i : time step

Accordingly, a multiplicative shift can be formulated as:

$$\dot{x}_i = x_i \cdot \frac{\hat{\bar{X}}_{fut}}{\hat{\bar{X}}_{ref}} \tag{1.5}$$

Analogous to eq.1.3, the delta-change concept can be extended to consider climate time series variance (modified from Ho et al. 2012):

$$\dot{x}_i = \mu_{\hat{X}_{fut}} + \frac{\sigma_{\hat{X}_{fut}}}{\sigma_{\hat{X}_{ref}}} \cdot (x_i - \mu_{\hat{X}_{ref}}) \quad \text{with} \tag{1.6}$$

$\dot{x}_i$: perturbed ("calibrated") climate variable at time step i

x_i : measured climate variable at time step i

$\mu_{\hat{X}_{fut}}$: location parameter of future simulated climate time series (e.g. mean)

$\mu_{\hat{X}_{ref}}$: location parameter of reference period climate time series (e.g. mean)

$\sigma_{\hat{X}_{fut}}$: scale parameter of future simulated climate time series (e.g. standard deviation)

$\sigma_{\hat{X}_{ref}}$: scale parameter of reference period simulated climate time series (e.g. standard deviation)

Note that subtle differences exist between eq.1.3 and eq.1.6, as the former starts with the variability of the climate model, whereas the latter starts with the variability of the observations (Hawkins et al. 2013). The delta-change approach has been extensively described, applied and compared (Ines and Hansen 2006; Lenderink et al. 2007; Räisänen and Ruokolainen 2008; van Roosmalen et al. 2010; Berg et al. 2012; Ho et al. 2012; Kling et al. 2012; Rasmussen et al. 2012; Teutschbein and Seibert 2012; Watanabe et al. 2012; Hawkins et al. 2013; Seaby et al. 2013). Nonetheless, the approach cannot apprehend for the shape of the distribution, since all events are adjusted with the same factor. Adding to this, the approach cannot be used to adjust frequencies, e.g. in wet and dry days. Therefore this method is used mainly for climate variables at coarse time resolution (e.g. monthly mean) or if distributions/thresholds are negligible.

Very similar to DC, "Linear scaling" (LS) is described in literature (Lenderink et al. 2007), which is the same as the straightforward bias correction (eq. 1.2) and has been applied also on future time series (Teutschbein and Seibert 2012): $\dot{x}_i = \hat{x}_{fut_i} - \bar{q}_{ref}$. Variations exist, as Lenderink et al. (2007) first corrected $\hat{X}_{ref}$ to obtain $\dot{X}_{ref}$, which was subsequently used for correction of future time series $\dot{X}_{fut} = \dot{X}_{ref} + (\bar{X}_{fut} - \bar{X}_{ref})$. Furthermore, Schmidli et al. (2006) extended the DC-approach for adjustment of wet- and dry-day frequency, referred to as "Local intensity scaling" (LOCI). More variants are given by tab. 1.4.

Distribution-based correction has been described by Ines and Hansen (2006) and Piani et al. (2010). This so-called quantile mapping (QM) maps the cumulative distribution functions (cdf) of both χ and $\hat{\chi}$, deriving the transfer function for correction. Hereby fitted distribution functions depend on the climate variable. Hence, normal distribution and two-parameter gamma distribution are respectively used for temperature and precipitation (e.g. Ines and Hansen 2006; Piani et al. 2010; Haerter et al. 2011; Vujadinović et al. 2012). Gamma distribution may also be used to correct solar radiation (Baigorria et al. 2008; Mudelsee et al. 2010). The latter is given by:

$$\begin{aligned}
cdf_\gamma(x, k, \theta) &= \int_0^x pdf(x, k, \theta)dx + cdf(0) \quad \text{with} \\
pdf_\gamma(x, k, \theta) &= x^{k-1}\frac{e^{-\frac{x}{\theta}}}{\Gamma(k)\theta^k}; \quad x > 0; \; k, \theta > 0 \\
\Gamma(k) &= \int_0^\infty e^{-t}t^{k-1}dt
\end{aligned}$$ (1.7)

$cdf_\gamma(x, k, \theta)$: value of the cumulative gamma distribution function

$cdf(0)$: fraction of days with no precipitation

$pdf_\gamma(x, k, \theta)$: value of the probability density function (gamma distribution)

x : value for which pdf and cdf are calculated (any possible value of a climate variable x)

k : form parameter

θ : scaling parameter

$\Gamma(k)$: function value of the gamma function

However, goodness of fit is not always given and non-parametric methods have been proposed as well (Piani et al. 2010; Hoffmann et al. 2012). Hereby values of the probability density function (pdf) can be estimated non-parametrically (pdf_{kernel}) with the help of kernel density estimation, applying a gaussian kernel:

$$pdf_{kernel}(x,h) = \sum_{i=1}^{n} \frac{1}{nh\sqrt{2\pi}} e^{-\frac{(x-\beta_i)^2}{2h^2}} \qquad (1.8)$$

pdf_{kernel} : probability density function value over all time steps i

β_i : climate variable at time step i

x : any possible value of β

h : bandwidth of kernel smoothing window

n : number of elements of β

i : time step

The smoothness of the kernel density estimate relies heavily on the choice of bandwidth h. The latter can be optimized by leave-on-out cross validation as described by Brooks and Marron (1991) and illustrated by Mudelsee et al. (2004).

Having estimated the individual cdf of each measured and simulated time series by calculation of $cdf(x)$ and $cdf(\hat{x})$ for all x and $\hat{x}$, a transfer function $\dot{X} = f(\hat{X})$ can be constructed. Applying a gamma distribution, the transfer function is (modified from Piani et al. 2010 and Teutschbein and Seibert 2012):

$$\dot{x}_i = f_\gamma^{-1}\Big(f_\gamma(\hat{x}_i, \hat{k}_{ref}, \hat{\theta}_{ref}), k, \theta\Big) \quad \text{with} \qquad (1.9)$$

f_γ : Gamma cdf (f_γ)

f_γ^{-1} : inverse of the Gamma cdf (f_γ)

Besides linear and distribution based bias correction, various approaches use different correction or transfer functions or combinations of these. For instance, Bordoy and Burlando (2013) found improvement of RCM temperature and precipitation time series after applying a "Power Transformation" (PT) (Leander and Buishand 2007; Teutschbein and Seibert 2012):

$$\dot{X} = a \cdot \hat{X}^b \quad \text{with} \qquad (1.10)$$

a, b = parameters, estimated for each climate variable and model grid point

More variants of bias correction approaches are given by tab.1.4. However, the different bias correction (bc) procedures add to the complexity and uncertainty of the climate impact modeling chain, hindering simulation comparability. As bc-methods are applied on different climate variables, models, varying spatial as well as time resolution, this cumbersomeness is further increased. Hence the corresponding behavior and influence on the climate/climate impact signal has been investigated (tab.1.4). Hereby the following assumptions are made when applying bias correction (from Ehret et al. 2012).

1) Reliability   The climate model can project climate change effects, despite its error
2) Effectiveness   Bias is corrected without introducing side effects/new biases
3) Time invariance The bias correction method is valid for the future (parameters, transfer functions are time invariant)
4) Completeness   The reference period must cover the full range of the climate variable

Thereby the bias correction method can be partially validated, e.g. by applying the method on different non-overlapping time slices (Piani et al. 2010). For correction of future time series the transfer functions are derived using the full information of the past, preferably from time series longer than 40 years (Chen et al. 2011). In order to reproduce the intra-annual pattern, bc is applied separately for each month ("cascade bias correction", Haerter et al. 2011): For a 10 a time series of January, for a 10 a time series of February and so on. Regarding uncertainty, the choice of decades from which bias correction parameters are derived is less important, as uncertainty from the choice of GCM or SRES-emission scenario is larger (Chen et al. 2011). Nevertheless, the uncertainty about stationarity (see above assumption no. 3) can be considered as a major drawback of bias correction (Teutschbein and Seibert 2012). Finally, concerning bias correction in the narrow sense (eq. 1.2,1.3) and the DC-approach, the choice of the reference period from which bias correction parameters are drawn affects the variance of the bias corrected time series differently, as the variance from the DC-approach is controlled by the historical climate, whereas the variance from direct bias correction is controlled by the climate of the climate model (Rasmussen et al. 2012; Hawkins et al. 2013).

Bias correction is heavily discussed and concepts of climate model bias per se and correction of time series for the use with impact models are often imprecise. For instance, spatial or temporal offsets may be recognized as bias (Haerter et al. 2011 as quoted in Ehret et al. 2012). Ongoing discussions exist on the influence of the bc on the climate change signal (Giorgi and Coppola 2010) and its justification (Ehret et al. 2012). Hereby bc might alter the climate change signal if low values are corrected differently than high values or if the distribution of the values changes over time (Hagemann et al. 2011). Ehret et al. (2012) further criticize, that a bias correction "neglects(s) feedback mechanisms" and destroys all physical relationship between climate variables. For instance, the spatio-temporal covariance structure of a GCM/RCM field is altered. This lack of consistency due to separate bias correction of single climate variables may lead also to unrealistic results in impact studies (Hoffmann and Rath 2011). However, two dimensional bias correction can be reached through segmentation, as described by Piani and Haerter (2012)[a]. Hereby one climate variable is corrected for segments of a given range of the second climate variable (e.g. bias correction of precipitation for each quantile of temperature). A different solution is given by Ehret et al. (2012), through correction of the impact model outcome (or model adjustment [author's note]). This would however require a larger effort, since more impact models than climate runs exist, as well as additionally decrease the comparability of the results. Finally, Ehret et al. (2012) expect a low acceptance of this approach.

[a] Received 10.09.2012; Accepted 14.09.2012; Published 16.10.2012. The present work includes one publication on consistent or 2d bias correction published earlier (Hoffmann and Rath 2012b: Received 28.06.2011; Accepted 22.02.2011; Published 14.03.2012)

Regardless of its physical justification, bias correction has become a standard procedure in climate change impact studies (Ehret et al. 2012). Albeit it "adds significantly to uncertainties in modelling climate change impacts" (Teutschbein and Seibert 2012), its importance has been emphasized (Teutschbein and Seibert 2012). This is due to the fact, that most impact models deliver unrealistic values when used with raw simulated climate time series. In these cases, bc is indispensable (Chen et al. 2011). For example, bias correction of T, P and ETp is necessary to obtain realistic discharges (Lenderink et al. 2007). Hence, depending on the sensitivity of the impact model, the use of raw time series should be avoided (Hawkins et al. 2013).

Different methods of bias correction have been suggested depending on the purposes. The choice depends on the wanted time resolution and the statistical properties of the target parameter. For example, Lenderink et al. (2007) found similar responses of the annual discharge after direct use of raw simulated climate time series and after DC, stating that both methods are plausible to produce future climate. However, while LS or DC can correct for mean values, distribution sensitive parameters (thresholds, extreme values, higher statistical moments) must be addressed by a distribution based approach, e.g. QM. For instance, linear scaling of daily precipitation led to an underestimation of distribution quantiles, subsequently leading to the underestimation of "the occurrence of extreme river flow". (Leander and Buishand 2007). Therefore relative performances of various bias correction approaches are listed by tab.1.4.

Table 1.4: Bias correction methods (Resolution: H: hourly, D: daily, M: monthly, S: season, Y: year, V: various)

Method	Climate[a] Variable	Res.	Evaluation Variable	Res.	Performance[b]	Reference
LS	T, P	M	river runoff	M	no improvement	Kling et al. (2012)
DC	T, P	M	T, P	M	improvement only in T	Räisänen and Ruokolainen (2008)
BC^c/QM	P	D	P	M,S	BC^c>QM	Ines and Hansen (2006)
BC^c/QM	P	D	P frequen., intensity	M	QM>BC^c	Ines and Hansen (2006)
BC^c/DC	T_{max}	D	T_{max}[d]	D	DC>BC^c >raw	Hawkins et al. (2013)
LS/LOCI/PT/QM	T, P	D	streamflow	M	improved by all	Teutschbein and Seibert (2012)
LS/LOCI/PT/QM	T, P	D	flood peak	S	QM>PT>LOCI>LS	Teutschbein and Seibert (2012)
QM	T, T_{min}, T_{max}, P	D	river runoff	Y	QM>raw	Hagemann et al. (2011)
DC/QM	T, P	D	irrigation	Y	QM>DC	Rasmussen et al. (2012)
DC/QM	T, P	D	T, P, ETp	M,S	QM>DC	Seaby et al. (2013)
DC/PT/QM/QM*[d]/ HE[e]/EQM[f]/QM**[g]	T, P	M	T, P	M	—[h]	Watanabe et al. (2012)
LS/LS*[i]/HE[e]	T, P	D	T, P	V	improved mean by all, higher moments depend on criteria	Berg et al. (2012)
QM (variations)	P	D	P	D	problems with extreme values by all	Gutjahr and Heinemann (2013)
QM (1d)	T, P, R_G	H	$T, P, R_G,$ april freezes	M	QM>raw	(Hoffmann and Rath 2011)
QM (1d)	P, R_G	H	fraction R_{dif}	M	QM<raw	(Hoffmann and Rath 2011)
HE[e] (1d, 2d)	T, P	H	T-P-copula	H	2d>1d	Piani and Haerter (2012)

[a] Simulated uncorrected climate variables used for bc. Additional climate variables used but not corrected are not listed.
[b] Order of performance from comparison of methods is indicated by higher>lower
[c] BC: Bias correction in the narrow sense, (eq. 1.2,1.3)
[d] Q*: QM with non-parametric estimation of cdf
[e] Histogram Equalisation: Transfer function from fit to sorted $\hat{X}$ plotted against sorted X (Berg et al. 2012; Piani and Haerter 2012)
[f] EQM: Equidistant QM (Li et al. 2010)
[g] QM**: Variation of QM by Watanabe et al. (2012)
[h] No clear ranking, authors proposed to apply multiple bias-correction methods
[i] LS*: LS variation correcting for frequency of wet and dry days

1.2.7 Aggregation and interpolation

Climate models operate on different grids or grids with grid points which differ from the locations of weather stations. Hereby climate model spatial resolution is a key factor for climate impact studies, as impact models usually operate on a higher resolution. The dependency of the spatial resolution on climate model variables is illustrated by fig.1.4 for elevation. Hereby orographic variance is lost with increasing area. Therefore distance, area and area elevation from records have to be taken into account in order to compare grids or grids and station records. This has consequences for calculating biases, total amounts (e.g. catchment) or energy balances. As mentioned above, these offsets may be recognized wrongly as model bias (Haerter et al. 2011 as quoted in Ehret et al. 2012) when comparing simulated and measured time series. For instance, annual precipitation sums of station record and model grid point(s) may differ largely (Hoffmann and Rath 2011). Thus records are corrected for elevation and/or spatially interpolated. Correction of air temperature (2m above ground) for elevation can be achieved by applying the "standard environmental lapse rate" (Bordoy and Burlando 2013) of -6.4 K/1000 m to -6.5 K/1000 m (MPI 2006; You et al. 2008; Hoffmann and Rath 2011; Bordoy and Burlando 2013), which might however depend on site specific orography, considered area size and season (Rolland 2003).

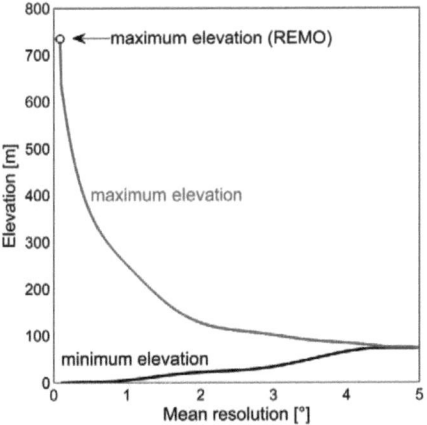

Figure 1.4: Influence of the choice of horizontal resolution on minimum and maximum area elevation. Minimum and maximum area elevations are shown for Lower Saxony, Germany, calculated from elevations of the regional climate model REMO. Maximum measured elevation: 971 m (Wurmberg).

It is questionable, whether simulated time series can be "bias corrected" or "calibrated" with the help from single weather stations. Hereby calibration of simulated time series for impact studies seems possible, if the regional representativeness of the time series is taken into account. The latter is subject to the regional orography and can be analyzed by means of semi-variance ("range") and variograms (Janis and Robeson 2004). However, despite a strong spatial dependance of some climate variables of orographically homogenous areas (e.g. radiation or temperature), it is suggested to use larger areas as well as multiple climate model grid points for comparison (MPI 2006). This is further due to findings of varying long-term trends in time series, even for "relatively homogenous area" (Pielke et al. 2000). Hence, although studies based on single climate model grid points exist (Bordoy and Burlando 2013), gridded data-sets are largely applied. For this purpose spatial interpolation ("gridding") is used. Several methods for interpolation of values for a wanted grid point exist, depending on the climate variable and orography. Basic interpolation methods are averaging, spline interpolation, inverse distance weighting and kriging. A different approach for generation of gridded data is the use of weather generators (Baigorria et al. 2008). Bordoy and Burlando (2013) chose to average daily temperature and precipitation of weather stations within a 20 km radius of a region with complex orography. This procedure was used for the sake of general validity and in order to avoid over-weighting of stations possibly influenced by microclimates. On the other hand, inverse distance weighting (IDW) gives more weight to closer records and might be suitable, if the spatial dependence of the climate variable at hand is known to decrease with distance. In its linear form IDW can be written as:

$$B' = \sum_{s}^{n}(b_s \cdot w_s) \quad \text{with} \tag{1.11}$$

$$w_s = \frac{u_s}{\sum_{i}^{n} u_i}$$

$$u_s = \frac{\sum_{i}^{n} a_i}{a_s}$$

$$A = a_i, .., a_n$$

$$B = b_i, .., b_n$$

A : Array of weighting criteria a, e.g. distances to station [km]

B : Array of records b to be weighted

B' : weighted Array

Hereby records (e.g. grid point time series or station record) are weighted inversely proportional to the distance. The procedure applies also to the weighting of model ensemble members, inversely to the model error.

Finally, geostatistic techniques as kriging are widely applied since they do take the spatial variance into account. In brief, kriging consists of generation of the experimental variogram, fitting of a variogram model and kriging interpolation. The method has the advantage of not being altered through clustering of samples, as weighting takes place. As several variants of the method exist (e.g. simple kriging, ordinary kriging, co-kriging, universal kriging, disjunctive kriging, indicator-kriging, multiple-indicator-kriging), literature is referred to (Oliver and Webster 1990; Gebbers 2010).

1.2.8 Uncertainties in climate impact projections

Mathematical models are simplifications of complexer systems which they intend to describe. Hence, deviations of simulated results and measurements/observations from the described system are to be expected. Hereby deterministic models (epistemic uncertainty: same input gives same output) and stochastic models (aleatory uncertainty: same input gives varying output) can be evaluated by means of uncertainty and sensitivity analysis (Marino et al. 2008), e.g. types of Monte Carlo simulation or latin hypercube sampling. Nevertheless, in climate impact studies complex models are concatenated (using model output for a second model as input), bias corrected, and applied with a limited number of repeated runs due to computing capacity (dynamical models). Thus, large uncertainties arise during the climate impact assessment simulation chain. These are due to measurement errors, data processing, model structures and concatenation among others. Tab.1.5 gives an overview on types of uncertainty according to different authors. Consequently, such basic doubts exist as on the suitability of regional climate models (Kerr 2013) for impact studies. Hereby, despite "huge uncertainties", present models were able to simulate the climate variability of the recent past (Førland et al. 2011; Maslin and Austin 2012; Maslin 2013) and their continuous improvement has been documented (Reichler and Kim 2008).

In order to estimate this "cascading uncertainty" (Maslin 2013), different approaches are used. Katz (2002) describes sensitivity, scenario and Monte Carlo analysis, where "scenario analysis is the technique most relied on". The latter can be used further, to identify the course of uncertainty of the different sources through variance decomposition (Hawkins and Sutton 2009; Yip et al. 2011; Hawkins and Sutton 2012). Hereby calculation of weighted variances across scenarios and across models allows the decomposition into model and scenario uncertainty as well as the estimation of the internal variability (residual variance). Olesen et al. (2007) additionally applied different impact models, regional and continental scales in order to quantify these sources of uncertainty. Furthermore, Bayesian approaches are increasingly used. For instance, Gouache et al. (2013) used 5 GCMs and different downscaling procedures for parameter estimation, obtaining a posterior distribution for 17 model parameters and for the variance of the residual error for the climate change impact on *Septoria tritici* blotch.

Table 1.5: Examples of sources of uncertainties in climate projections

Source	Description	Reference
Measurement error	Random error and bias	Katz (2002)
Sampling	Averaging over a finite number of years	Déqué et al. (2007)
Radiation	Uncertainty from emission scenario	Déqué et al. (2007)
Boundary	Uncertainty from GCM (In the case of RCM)	Déqué et al. (2007)
Variability	Systematic differences (space, time), e.g. spatial variability of precipitation	Katz (2002); Déqué et al. (2007)
Model structure	Model functions or relationship	Katz (2002); Solomon et al. (2007)
Scaling/aggregation	Model or data scale or aggregation	Katz (2002)

1.2.9 Observed and projected climate change

Despite large differences in the quality, length, selected time scale, completeness, distribution of stations and measuring methods of tracking climate by the different meteorological services as well as differences in climate model projection set-ups, estimates on different scales have been given (table 1.6). General trends are increasing mean temperature, decreasing number of frost days as well as an extension of the period of vegetation. However, differences in season, subareas and regarded time slice must be kept in mind as these might differ from the main trend, regarding direction and speed of the climatic change. For instance, in 2007 the IPCC Working Group I reported, that the last 50 years exhibited a global temperature increase of 0.13 °C $\pm$ 0.03 °C , being "nearly twice that for the last 100 years". Seasonal trends of precipitation in Germany and Lower Saxony are in the main positive for winter and negative for summer.

Future estimates depend largely on the chosen SRES emission scenario. Global estimates of air temperature increase for the end of the 21st century (2090-2099 compared to 1980-1999) given by Solomon et al. (2007) show the following range for the best estimator and likely range; in °C : B1 1.8 (1.1-2.9), A1T 2.4 (1.4-3.8), B2 2.4 (1.4-3.8), A1B 2.8 (1.7-4.4), A2 3.4 (2.0-5.4), A1F1 4.0 (2.4-6.4)

Table 1.6: Global and regional observed and projected changes in air temperature T and precipitation P. Values depict the annual mean if not further indicated. W: World, DJF: Winter, MAM: Spring, JJA: Summer, SON: Autumn; Regions are abbreviated by standard code (ISO-3166-1 Alpha-2).

Domain	Observed trend	Projected (End of 21st century)	Reference
W T	+0.7 ±0.2 °C, 1906-2005	+1.7 to +4.4 °C [a]	Solomon et al. (2007)
W P	-7 to +2 (mean: -3.7) mm/(10 a), 1951-2005	—	Solomon et al. (2007)
EU T	+1 °C, 1906-2005[b]	south +3.2 °C, north +3.5 °C [a]	Solomon et al. (2007)
EU P	north +6 to +8 %, 1900-2005	south -12 %, north +9 %[a]	Solomon et al. (2007)
DE T	+0.8 to +1.1 °C, 1901-2000	—	Schonwiese and Janoschitz (2008) as quoted in (Haberlandt et al. 2010)
DE T	+0.97 °C, 1901-2008	—	DWD (2009)
DE T	—	+2.5 to +3.5 °C [c]	UBA/MPI (2006)
DE T	—	+2 to +4.5 °C (DJF), +1.3 to +5 °C (JJA)[de]	Jacob et al. (2012)
DE P	+9.7 %, 1901-2008	—	DWD (2009)
DE P	—	-4 to +30 % (DJF), -25 to +10 % (JJA)[def]	Jacob et al. (2012)
DE-NI[g] T	+1.2 °C, 1881-2009	+2.5 °C [d]	Moseley et al. (2012)
DE-NI[g] T	+1.3 °C, 1951-2005	—	Haberlandt et al. (2010)
DE-NI[g] T	+1.09 °C, 1901-2008	—	DWD (2009)
DE-NI[g] P	+11.8 %, 1901-2008 (+23.6 % DJF, +1.2 % JJA)	—	DWD (2009)
DE-NI[g] P	+9 %, 1951-2005 (+32 % DJF, -13 % JJA)	—	Haberlandt et al. (2010)
DE-NI[g] P	+15 %, 1881-2009 (+30 % DJF)	+11 to +18.5 % (SON, DJF, MAM), -10 % (JJA)[d]	Moseley et al. (2012)

[a] 2090-2099 compared to 1980-1999, scenario A1B
[b] value extracted from figure SPM.4
[c] 2071-2100 compared to 1961-1990, range of scenarios A1B, B1, A2
[d] 2071-2100 compared to 1971-2000, scenario A1B
[e] Range over all presented climate realizations (ENSEMBLES, REMO)
[f] Most simulations showed a decrease of summer precipitation
[g] Present year-mean climate of Lower Saxony (1971-2000, Moseley et al. 2012): 8 to 9.5 °C air temperature at 2 m height above ground, 240-280 d mean duration of vegetation period, 50 to >80 days with frost, 5 to 6 m s^{-1} wind speed, 50 to 80 mm per month precipitation, 1 to 2 days with extreme precipitation in summer, 16 to 17 d longest dry period from April to September in most areas.

1.3 Susceptibility of plant / horticultural systems to climatic changes

1.3.1 Basic thoughts on climatic impact through changes in distribution parameters

The described projected changes of climate variables will have consequences for horticultural systems. These impacts can be deduced from changes in climate variable distributions concerning location, spread and shape. On the one hand changes in the location of a distribution (e.g. mean air temperature) can be meaningful due to the relative location to an optimum of a given biological process (e.g. effect of mean air temperature on cell doubling time). On the other hand, this shift will directly affect the appearance (number) and extent (scale) of extreme events. The same effect can be reached through changes in the shape of the distribution as shown by fig.1.5 for a normal distribution. The following examples can be given: Assuming normal distribution, an increase of mean air temperature has led to a longer vegetation period as well as earlier flowering of temperate fruit trees. However, this does not necessarily reduce cold stress or late frost risk due to a simultaneous increase in the variance of temperature. Depending on the latter, only little changes might occur in minimum temperatures (lower edge of the distribution). Quantification of this matter is subject of present research and further influences are described in the following sections. A second example is the energy demand of greenhouse heating systems, which can be expected to decrease in the mean with ongoing global warming. Nonetheless, dimensioning of the heating systems must take lowest rather than mean temperatures into account.

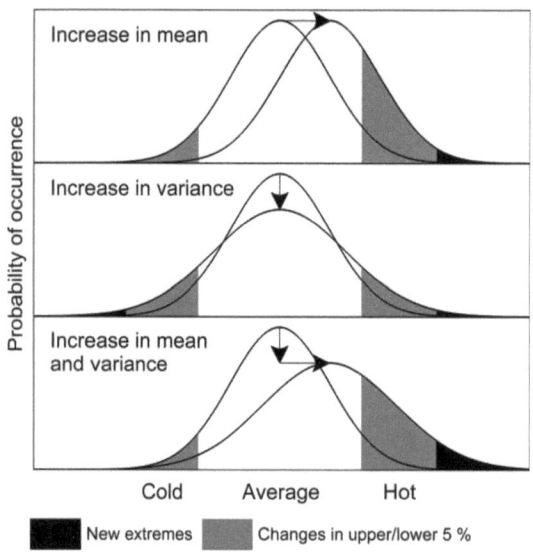

Figure 1.5: Basic concept of changes in distribution mean and variance for normally distributed climate variables. Modified from Houghton et al. (2001), chapter 2.7.1, fig. 2.32.

1.3.2 General system parameters

Plant biology is subject to climatic changes at all organizational levels, ranging from gene expression, single cells, organs and finally plants up to plant communities and ecosystems (Morison and Morecroft 2006). Hereby single climate variables may have several effects on plant physiology, which may be counteracting or may not necessarily be directly related to plant growth or development. Therefore the effects of one single climate variable integrate to the effect of that climate variable on total metabolism (e.g. temperature, Porter and Semenov 2005), whereas the effects of all climate variables integrate to the climatic impact. In order to comprehend a possible range of climatic impacts on horticultural production systems, an overview of main physiological responses is given first, followed by considerations of observed and expected future climatic impacts at the production ("yield") level. It must be remarked further, that effects at one level (e.g. response of photosynthetic fixation rate to carbon dioxide concentration) may also become insignificant at a higher level of abstraction (e.g. C_4 plants do not benefit from increased atmospheric carbon dioxide concentration as C_3 plants). Concerning horticultural production systems, the following considerations are restricted to the physiological-to-plant levels of vascular plants. While Santos et al. (2012) differentiates between climate variables as forcing factors and as extreme events, in the present work these are treated as effects of different quantiles of a distribution of the same climate variable(s). Furthermore, large interactions exist between climate and plant pests or pathogens or weeds. These fields constitute major research fields on their own and hence literature is referred to (Woiwod 1997; Harvell et al. 2002; Fuhrer 2003; Juroszek and v. Tiedemann 2011, 2012).

Carbon dioxide
Plant dry matter consists to about 40 to 45 % of carbon (Krug et al. 2002), which is allocated from the atmosphere to the plant through photosynthesis (≈ 120 Gt a^{-1} global uptake, Bowes 1991). Hence atmospheric carbon dioxide concentration affects the photosynthetic rate, as it is substrate to Ribulose-1,5-bisphosphate carboxylase oxygenase (RuBisCO) (Taiz and Zeiger 2000; Campbell and Reece 2003). Assuming a dependency of internal leaf CO_2 concentration on atmospheric CO_2 concentration (Katul et al. 2000), an increasing CO_2 concentration enhances the CO_2 fixation rate to a certain extent, as shown for example by Wollenweber et al. (2003) for light saturated leaf photosynthesis in response to internal leaf CO_2 concentration. However, C_4- and CAM-plants raise the CO_2 concentration at the site of photosynthesis through 'CO_2 pumping' (Taiz and Zeiger 2000: 209-214) and storage during the night (Taiz and Zeiger 2000: 214-216), respectively. Depending on CO_2-feedback (e.g. starch accumulation), this translates into the respective growth responses of plants (Bowes 1991). For example, Quebedeaux and Chollet (1977) grew plants of C_3 and C_4 *Panicum* species in a "controlled environment growth room" under varying CO_2 and O_2 concentrations. Hereby CO_2-enrichment clearly increased the growth of C_3-plants, contrary to C_4-plants. Additionally, numerous experiments have been carried out in order to identify the influences of environmental CO_2 enrichment on stomatal conductance, transpiration and water use efficiency (WUE). In the main, stomata conductance and transpiration are decreased (Kimball and Idso 1983; Rötter and Van De Geijn 1999),

whereas WUE is increased (see Bowes 1991; Rötter and Van De Geijn 1999 for a review). Rötter and Van De Geijn (1999) further discuss a possibly reduced stress from NO_x and O_3 due to CO_2 -stomata regulation. However, these effects have to be regarded with care at the plant level, since CO_2 may stimulate leaf area (see Morison and Lawlor 1999 for a review). Further anatomical, morphological, physiological or biochemical influences as well as effects of adaptation to increased CO_2 have been described (see Bowes 1991 and Moretti et al. 2010 for a review). Interactions with root activity and soil micro-biota exist, as the diffusion of CO_2 in soils depends on atmospheric CO_2 concentration.

Summarizing, while global atmospheric CO_2 concentration has been rising continuously during the recent past, reaching a current level of 394 ppm (2012 annual mean at Mauna Loa, Hawai, CO_2 expressed as a mole fraction in dry air, μmol mol^{-1}, abbreviated as ppm, Tans 2013), a doubling in CO_2-concentration could increase yield by about 33 %, as concluded by Kimball and Idso (1983) from a review of 430 experiments. Although the latter result has been confirmed for *Triticum aestivum* L. by Amthor (2001) (156 experiments, including FACE experiments) with 31 % increased yield by doubling of CO_2 from 350 to 700 ppm, uncertainty remains as other sources report an about 50 % lower expected increase in yield, when regarding FACE-experiments (Long et al. 2006). Concludingly, the 4th Assessment Report of the IPCC reported a positive effect of CO_2 alone on yield (Parry et al. 2007), whereas combined effects of CO_2 and climate on crop growth and yield remain limited (Ewert 2004).

Temperature

As plants are poikilothermic, their metabolism is tied to the temperature of the surrounding medium (Körner 2006). Since enzyme kinetics are temperature-dependent (Campbell and Reece 2003), temperature exerts multiple influences on physiological processes. Main influences exist on net photosynthesis as respiration (mitochondrial respiration as well as photorespiration) increases with temperature (Taiz and Zeiger 2000; Sage et al. 2008). Besides respiration, possible effects limiting the photosynthetic capacity above leaf optimum temperature are discussed (Sage et al. 2008). Additionally temperature affects photosynthesis by diffusion limitation through greater stomatal limitations at higher temperatures concerning vapor pressure deficit (vpd) (large vpd can reduce stomatal conductance) and CO_2 -response of Rubisco-limited/RuBPregeneration-limited CO_2 -assimilation rate as well as through limitation of mesophyll transfer conductance (Sage et al. 2008).

Further, temperature strongly influences the cell-doubling time (Fig. 1.6, Körner 2003). Hence, under optimal conditions temperature determines the velocity of plant development (Krug et al. 2002: 34). For instance, Porter and Semenov (2005) state a linear relation between the mean phase temperature and the rate of plant development in general. Hereby plant susceptibility to temperature varies largely and plant species can be grouped according to their range and optimum (cardinal temperatures). Recapitulating, the order of the temperature-limits for plant processes can be given from wide to narrow as Lethal > Activity > Development > Growth > Reproduction (Porter and Semenov 2005). From this, mean, extremes and variance of temperature determine plant growth/development. Hence in temperate climates, the vegetation period is limited by low temperatures (winter); (see Bender and Schaller (2012) for definitions of vegetation period). Frosts or low temperatures cause damages up to death, depending on plant susceptibility and phenological stage (Friedrich and Fischer 2000: 243-253; Krug et al. 2002: 37; Jackson 2003: 282-286). In the case of blossom frost, reproductive organs may dry out due to cuticular cracks after tissue freezing (Rodrigo 2000). On the other hand, heat may lead to death, damages, bolting or deficiencies regarding quality (Krug et al. 2002: 37) or decreased fertility (Wollenweber et al. 2003). In the latter case, heat stress during anthesis or double-ridge stage has led to reduced yield (Wollenweber et al. 2003). Hereby effects of mean temperature increases must be differentiated from changes in temperature variability. For instance, in a warming world blossom frost occurrences could increase due to faster advancing of flowering dates than dates of the last spring freeze (effect of mean temperatures). However they could also increase due to increased variance of either flowering dates or last dates of the last spring freeze (effect of increased variability) (Chitu et al. 2012).

In the temperature range between lower and upper limit to damages, plants show short-term responses, e.g. heat-shock protein synthesis (Taiz and Zeiger 2000: 736-741) or altered root hydraulic conductance (Taiz and Zeiger 2000: 84) as well as long-term responses to changes of temperature. Altered metabolism including stress responses may appear on both scales. For example, seedlings of *Lactuca sativa* show increased chicoric and chlorogenic acid and accumulate Quercetin-3-O-glucoside and luteolin-7-O-glucoside, while leafs and roots of *Brassica oleraceae* show increased levels of aquaporins and glucosinolates as well as of the fatty acids linoleic and stigmasterol, but decreased levels of the fatty acids palmitoleic, oleic and sistosterol (Ahuja et al. 2010). Anthocyanin accumulation in pomegranate (*Punica granatum* L.) is influenced by seasonal temperature, hence determining the color of the fruit (Borochov-Neori et al. 2011). Regarding climate, long-term responses as altered CO_2 assimilation rate and C and N partitioning (Morison and Lawlor 1999; Porter and Semenov 2005), chilling requirement (Luedeling et al. 2009a; Luedeling 2012) or vernalization (Wurr et al. 1996; Taiz and Zeiger 2000: 707-708; Krug et al. 2002: 88-91) may become relevant. Regarding the latter two, low temperatures serves as a signal (Körner 2006: 61). In the case of fruit tree phenology, phenological timing in spring (e.g. bud break, flowering) is determined by the temperature dependent processes chilling and forcing (Legave et al. 2008b,a).

Regarding horticultural production, air temperature has numerous further effects on the production chain, e.g. effects on greenhouse energy demand (Krug et al. 2002: 120, Hoffmann and Rath 2009) with regard to heating or cooling systems, on the possible period of seed storage (Krug et al. 2002: 215), the quality (reduced sugar content under high temperature in pea, Abou-Hussein 2012), fruit set (reduced fruit set caused by warm temperatures during bloom; sweet cherry: Hedhly et al. 2007; tomato: Sato et al. 2006) or on the shelf-life of harvested products (Taiz and Zeiger 2000: 310; Krug et al. 2002: 258; Moretti et al. 2010). However, innumerable interactions with other climate variables, e.g. with CO_2 (Morison and Lawlor 1999), as well as with the plant-soil interface (rhizosphere), e.g. ion chemistry, micro-biota or soil water content through evapotranspiration, exist. Finally, yield response to temperature varies regionally, as it is limited in northern Europe by cool temperatures (Holmer 2008), whereas in southern Europe high temperatures in combination with low precipitation are limiting (Reidsma and Ewert 2008).

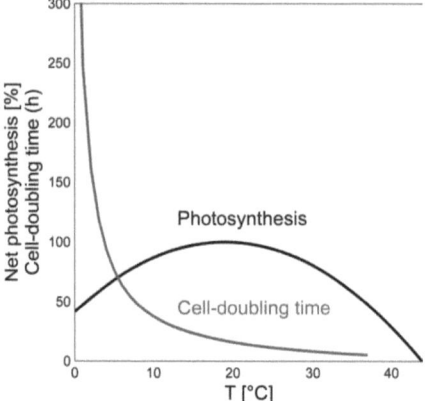

Figure 1.6: Temperature influence on leaf net photosynthesis and cell cycle. Modified from Körner (2006) and Körner (2003).

Precipitation

Precipitation as climate variable (see Bender and Schaller 2012 for definitions of precipitation indexes) affects plant development and growth indirectly via the available soil water. Hereby precipitation adds to soil water content as follows (modified from Krug et al. 2002: 39; in mm):

$$\Delta W = P + C - T - E - R - D - I + IRR \quad (1.12)$$

ΔW : Difference in soil water content
P : Precipitation
C : Capillary rise from ground water
T : Transpiration
E : Evaporation
R : surface runoff
D : Drainage to groundwater
I : Interception
IRR : Irrigation

Depending on the present soil, the soil water content determines the soil water tension (Scheffer and Schachtschabel 1989: 178) and hence hydraulic conductivity of the soil (Scheffer and Schachtschabel

1989: 184; Taiz and Zeiger 2000: 82). Hereby the rate of root water uptake, e.g. responsible for nutrient transport and plant turgor, depends on the hydraulic conductivity of the soil and interacts with evapotranspiration. Therefore increasing soil water tension eventually decreases plant water uptake, as can be verified with the help of relative transpiration rates ($ETa\ ETp^{-1}$, Scheffer and Schachtschabel 1989: 201). While the latter eventually decreases with increasing soil water tension, this effect is further enhanced through higher evapotranspiration rates (Scheffer and Schachtschabel 1989: 201). Hence, in crop production deficits in soil water content must be compensated through irrigation (capillary rise can be neglected). For the numerous effects of decreased water availability on crop production, literature must be referred to (Turner and Kramer 1980; Simpson 1981; Raper and Kramer 1983; Marschner 1995; Taiz and Zeiger 2000; Krug et al. 2002). Again, short- and long-term responses/adaptation take place (Krug et al. 2002: 197). Via remediation through irrigation, precipitation further influences horticultural crop production, as technical infrastructure is required for open field (irrigation systems) as well as greenhouse crop production (storage basins). Contrary to water deficit conditions, water logging may further occur, interacting with O_2 depletion/deficiency in soil and roots, soil erosion (Scheffer and Schachtschabel 1989: 389,470) as well as altered nutrient availability (Marschner 1995: 626-643). Further effects on horticultural production are due to precipitation in solid form (hail damages, field snow cover, greenhouse snow cover).

Shortwave radiation

Global radiation is the "total short-wave radiation from the sky falling onto a horizontal surface on the ground", including "both the direct solar radiation and the diffuse radiation resulting from reflected or scattered sunlight" (PIK 2013), with wavelengths ranging from 290 to 2800 nm (Krug et al. 2002: 30). Photosynthetic active radiation (PAR), constituting roughly 50 % of the global radiation (Krug et al. 2002: 30) is the driving force to photosynthesis (Campbell and Reece 2003: 209-228). Species show characteristic response curves to PAR, depending further on CO_2 and temperature (see Krug et al. 2002: 94 for an example for a leaf of *Cucumis sativus*). Together with temperature, global radiation determines evapotranspiration (Krug et al. 2002: 67) as well as greenhouse energy demand (Krug et al. 2007). Additionally, the circadian rhythm as well as flowering induction (photoperiodism) and possibly vernalization (Taiz and Zeiger 2000: 708-709) are influenced by global radiation (Taiz and Zeiger 2000: 699-707). Hereby photoperiodism constrains the influence of temperature on development (e.g. synchronization of flowering in populations or preventing too early flowering or too late induction of dormancy; Körner 2006). Stomata regulation is influenced by photon flux and light quality ("blue light") as shown for *Vicia faba* (Taiz and Zeiger 2000: 525-526), hence affecting transpiration.

Concerning yield, the given necessity of PAR for plant growth has to be put in context to other light quality criteria (e.g. latitude and/or length of the day, sine of the solar angle and fraction of diffuse radiation, level of UV-B radiation Schultz 2000) and other environmental influences. For instance, De Temmerman et al. (2002) surprisingly found decreasing tuber yield of *Solanum tuberosum* with increasing total radiation (emergence to harvest) on different (Europe) as well as same locations (e.g. Gießen, Germany). The authors ascribe this result to possible positive effects to increased day length in summer as well as to possibly associated temperature, senescence/ripening and drought effects.

Furthermore, quality degradation in vegetables has been described as a result of the combination of elevated temperature and solar radiation (Moretti et al. 2010). Hereby leaf injuries (Lettuce), discoloration (Apple, Avocado, Lime, Pineapple, Snap bean) or both (Cabbage) as well as sunburn (Muskmelon, Bell pepper, Tomato) are mentioned.

Other climate variables

Most climate impact studies focus on effects of the mentioned climate variables CO_2, temperature and precipitation as these are expected to have the strongest impact (both climatic change and effect on plant growth and yield). However further climatic influences as of wind speed (risk of windthrow) or of O_3 (toxicity, Booker et al. 2009; Fuhrer 2009; Moretti et al. 2010) can be found. Additionally effects of other climate variables as relative air humidity can be related to temperature, precipitation and global radiation or may only exert an effect in combination with other additional factors (e.g. wind direction in combination with exposure/orography). Depending on the definition of climate, extended or derived variables may also be regarded as "climate variable", e.g. river discharge, land cover, groundwater, albedo, fire disturbance or soil carbon (http://www.wmo.int/pages/prog/gcos/index.php?name=EssentialClimateVariables, lastfetch:24.03.2013). These may further influence plant growth and yield or depend themselvs on plant growth, e.g. leaf area index (LAI) or above-ground biomass.

1.3.3 Vulnerable systems

According to (Olesen et al. 2011), climate change influences crop production
a) Directly — CO_2 effects and "resource use efficiencies"
b) Directly — temperature, rainfall, radiation and others on crop development and growth
c) Indirectly — shifts in crop suitability (e.g. expansion to north)
d) Directly — damages due to extreme events
e) Indirectly — environmental pollution

Hereby corresponding literature may be biased, as most literature concentrates on a-c (Olesen et al. 2011). However, sensitivity of a horticultural system to climatic changes can be defined as its vulnerability ("the inability to withstand the effects of a hostile environment"; http://en.wikipedia.org/wiki/Vulnerability). While section 1.3.2 describes single climatic effects (e.g. on physiology), systems vulnerability comprises the system with regard to yield, multiple effects and/or other climate variables (Wand 2007; Wand et al. 2008). Reviews on the vulnerability of agriculture (Zebisch et al. 2005: 67-83; Schaller and Weigel 2007) and horticulture (Fink et al. 2009) have been published.

General statements can be found, e.g. that an increase of 2 °C will not affect most plants, but the accompanying precipitation/soil moisture will (Marris 2007). Santos et al. (2012) state, that C_4 and CAM plants "tend to be more tolerant to climate change than C_3 plants". However, in more detail and in a survey over 13 environmental zones in Europe, Olesen et al. (2011) (see also Olesen and Bindi 2002 for main climate influences in Europe) estimated the following effects: Late frost in spring or early frost during fall (Limiting factor), high frequency of rainy conditions (problematic at sowing or harvest), flooding/stagnant surface water (persistent problem for grasslands, winter wheat or spring barley), overwintering/damage to crops during winter (major problem for grasslands and winter wheat production in boreal/alpine north Europe), hail damages (no measurable effect on large scales with regional exceptions), drought (mainly not limiting although large concern, winter wheat limited across warm and cold regions, critical at flowering and seed setting), heat stress (considered important, limiting when coinciding with drought). In fruit production, climate change possibly overcomes "the adaptibility of many temperate fruit crops" (Campoy et al. 2011). For instance, temperate climate fruit tree production could further be affected by alteration of flowering synchrony (Campoy et al. 2011).

Regardless of systems limiting factors, several climatic influences can degrade crop quality and hence lead to a complete production loss or an unsalable product. For horticultural production, these so-called "Knock-Out-Effects" determine high vulnerability. Examples are the obligatory vernalization of cauliflower in order to initialize curd growth as well as the winter chill requirement of several fruit trees in order to break dormancy (Campoy et al. 2011). In these cases, the specific temperature requirement restricts crop production to a narrow temperature/time-frame. Hence, in these cases the present form of crop production could eventually cease completely due to changes in temperature beyond this frame (Luedeling et al. 2009b). Similarly, water deficiency can cause total loss in horticulture, as products of low quality are not marketable (Fink et al. 2009). This differs from agriculture, where water shortage decreases yield in the mean. An overview of Knock-Out-Effects for horticultural products is given by table 1.7, (Hoffmann and Rath 2010).

Table 1.7: Potential "Knock-Out-Effects" (Hoffmann and Rath 2010)

Variable	Factor	Culture	Measure
temperature	vernalization	*Brassica oleracea* var. *botr.* L., *Brassica oleracea* var. *gong.* L.	cultivation methods, breeding
	frost	*	sprinkling
	heat	*	shading, irrigation
	dormancy	fruit trees	chemical treatment
	energy costs	miscellaneous	cultivation methods
precipitation	waterlogging	*Brassica oleracea* var. *gong.* L.	drainage, covering
	drought	*	shading, irrigation
	heavy rain	*	covering
	fruit cracking	*Prunus avium*	possibly breeding in combination with shelters
	damage by hail	fruit cultivation	hail nets
	leaching	*	cultivation methods, fertilization
solar radiation	bolting	*Lactuca sativa* var. *cap.* L.	shading
	burn	*Malus domestica*	shading
	seediness	*Brassica oleracea* var. *botr.* L.	shading
wind	mechanical damages	fruit cultivation	wind screen, location
O_3, NMVOC	toxicity	*	—
miscellaneous	unwanted compounds	*	—

∗) applies to most open field cultures

1.3.4 Observed climatic impact

The described main effects of climate variables on plant development were derived from measurements / observations with varying natural or artificial climate conditions and not necessarily from long time series. However, the other way around observed long-term changes in these effects can be attributed to changes in climate parameters. This is certainly difficult, as most effects are masked by technological advances, breeding, economical situations and environmental interdependencies of variables. While it still has been done extensively for agricultural crops, knowledge about climate impacts on horticultural crops remains low (see Abou-Hussein 2012 for a review on climate change and vegetable crops). This is mainly due to the lack of large observational time series (>30 a). One exception is temperate climate fruit tree phenology, thanks to large observational systems (e.g. meteorological services) around the world.

Crop phenology since 1951 reveals almost entirely negative trends, implying that events occur earlier with increasing temperature (tab. 1.8). Estrella et al. (2007) examined phenological stages of 78 horticultural and agricultural crops, with 97 % of the records showing an advancing of phenophases. Exceptions are the delayed heading of *Z. mays* and harvest of *B. napus* var. *napus*, whereas for the date of fruit maturity of *Malus domestica* no clear signal was found, albeit significant earlier flowering. A delayed and irregular emergence of *Asparagus* was observed, possibly due to deficient winter chill (Fink et al. 2009).

Altered phenology consequently led to changes of grape acidity, sugar content and/or phenolics (Campeanu et al. 2012; Vršic and Vodovnik 2012), thus affecting product quality directly. Hereby decreasing acidity had positive (red wine, Campeanu et al. 2012) as well as negative (white wine, Campeanu et al. 2012) influences on wine quality.

Lobell and Field (2007) attributed differences in global yield (FAO data) to climate change by applying regression models on detrended time series by taking year-to-year differences (=first differences) and estimating the uncertainty (confidence interval) for each decade via bootstrapping. Although this simple approach certainly cannot explain cause-effect relations mechanistically, it contains two assumptions. First, climate variation causes variation of yield (attribution of cause-effect direction). The second assumption is, that first differences "minimize the influence of slowly changing factors, such as crop management". Hereby variances in yield changes could partially be explained with minimum and maximum air temperature as well as precipitation. As a result, despite increasing yields (t ha^{-1}), increasing temperatures showed a clear negative influence on the yield of wheat, maize and barley (1961-2002). Hereby significant yield effects were mostly temperature driven, while precipitation had "only minor influences". However, this signal was less pronounced for rice, soybean and sorghum. Furthermore, regarding global trends does not rule out local yield increases due to climate change, as the examples of Turkey (Ulukan 2009) as well as Greece (decrease) and Finland (increase) show (tab. 1.8). Detailed studies on heat stress do not exist for horticultural crops in general, but Fink et al. (2009) mention diminished quality of some vegetables due to night temperatures >20 °C and daytime temperatures >30 °C.

Table 1.8: Observed abiotic impact of climate change. W: World; Regions are abbreviated by standard code (ISO-3166-1 Alpha-2)

Factor	Species	Impact (-,∘,+)[a]	Reference
Begin of growing season (DE)	—	-	Chmielewski et al. (2004)
Sowing (DE)	Z. mays	-	Estrella et al. (2007)
Emergence (DE)	B. vulgaris, S. cereale, Z. mays	-	Chmielewski et al. (2004); Estrella et al. (2007)
Emergence (DE)	Asparagus	+[b]	Fink et al. (2009)
Begin of stem elongation (DE)	S. cereale	-	Chmielewski et al. (2004)
Heading (DE)	Z. mays	+	Estrella et al. (2007)
Harvest (DE)	B. vulgaris, S. cereale, Z. mays	-	Chmielewski et al. (2004); Estrella et al. (2007)
Flowering date	M. domestica[c]	-	Chmielewski et al. (2004); Wolfe et al. (2005); Estrella et al. (2007); Chmielewski et al. (2009); Blanke and Kunz (2009); Sugiura (2010); Kunz and Blanke (2011); Blanke and Kunz (2011)
Flowering date (DE)	P. avium	-	Chmielewski et al. (2004, 2009)
Flowering date (JP)	Cerasus sp.	-	Miller-Rushing et al. (2007)
Flowering date (JP)	Prunus sp.	-	Miller-Rushing et al. (2007)
Flowering date (US)	P. serrulata 'Kwanzan'	-	Chung et al. (2011)
Flowering date (US)	P. ×yedoensis 'Yoshino'	-	Chung et al. (2011)
Flowering date (DE)	Pyrus communis	-	Estrella et al. (2007); Chmielewski et al. (2009)
Flowering date (FR)	P. armeniaca	∘	Legave and Clauzel (2006)
Flowering date (DE)	P. armeniaca	-	Chmielewski et al. (2009)
Flowering date (DE)	P. domestica	-	Chmielewski et al. (2009)
Flowering date (DE)	P. persica	-	Chmielewski et al. (2009)
Flowering date (US)	S. chinensis	-	Wolfe et al. (2005)
Flowering date (US)	V. vinifera	-	Wolfe et al. (2005)
Flowering date (DE)	R. uva-crispa	-	Estrella et al. (2007)
Fruit ripe for picking (DE)	M. domestica	-,∘	Estrella et al. (2007); Chmielewski et al. (2009)
Fruit ripe for picking (DE)	P. cerasus	-	Estrella et al. (2007)
Fruit ripe for picking (DE)	R. fruticosus	∘	Estrella et al. (2007)

1.3. SUSCEPTIBILITY TO CLIMATIC CHANGES

Factor	Species	Impact (-,∘,+)[a]	Reference
Fruit ripe for picking (DE)	P. domestica	∘	Estrella et al. (2007)
Harvest (DE)	B. napus var. napus	+	Estrella et al. (2007)
Blossom frost risk (DE)	M. domestica[c]	∘	Blanke and Kunz (2009); Kunz and Blanke (2011); Blanke and Kunz (2011)
Freezing Injury (JP)	fruit trees	-,∘,+	Sugiura (2010)
Blossom frost risk (JP)	fruit trees	-,∘,+[d]	Sugiura (2010)
Grape acidity content	V. vinifera	-	Campeanu et al. (2012); Vršic and Vodovnik (2012)
Grape sugar content	V. vinifera	+	Campeanu et al. (2012)
"Absolute" maturity	V. vinifera[e]	-	Campeanu et al. (2012)
Berry crop bud break date	various	-	Kampuss et al. (2009)
Berry crop bud flowering date	various	-	Kampuss et al. (2009)
Berry crop frost risk	various	+	Kampuss et al. (2009)
Berry crop chill	various	-	Kampuss et al. (2009)
Berry crop bud break	various	-	Kampuss et al. (2009)
Woodlands	Juniperus L.	-	Fisher (1997)
Winter chill availability	—[f]	-	Luedeling (2012)
Winter chill availability	fruit/nut trees	-	Luedeling et al. (2009b)
O_3	—	+	Fuhrer (2009)
Yield (W)	T. aestivum, Z. mays, H. vulgare L.	-	Lobell and Field (2007)
Yield (Fl)	cereal, tuber crops	+	Olesen et al. (2011)
Yield (GR)	cereal, tuber crops	-	Olesen et al. (2011)

[a] decrease or advancing in DOY (-), no change (∘), increase or delay in DOY (+)
[b] Emergence was delayed and not uniform
[c] cvs 'Booskoop', 'Cox's Orange Pippin', 'Golden Delicious'
[d] mainly increasing in the North of Japan
[e] cv. Sauvignon, cv. Cabernet S., cv. Merlot, cv. Pinot Gris, cv. Riesling Italian
[f] Winter chill availability as calculated with various models

1.3.5 Expected future impact

The 4th Assessment Report (AR4) of the International Panel on Climate Change (IPCC) (Parry et al. 2007) concluded, that "climate change and variability will impact food, fiber, and forests around the world due to the effects on plant growth and yield of elevated CO_2, higher temperatures, altered precipitation and transpiration regimes, and increased frequency of extreme events, as well as modified weed, pest and pathogen pressure". While the authors state, that these findings have been confirmed by following studies, literature reports a wide range of possible future climatic impacts. Due to the lack of large-scale studies on horticultural crops and for reasons of comparison, the following review does include examples for agricultural field crops.

In the main, yields will respond depending on the regional climatic situation. Trnka et al. (2011) examined the agroclimatic conditions in Europe and found several cases of increasing drought stress and a shorter growing season "squeezed between a cold winter and a hot summer", stating a higher risk for rain-fed crops as well as a generally increasing inter-annual yield variability. Differently, northern latitudes were also found to benefit from larger total vegetation periods, increasing temperatures and CO_2 (Olesen and Bindi 2002; Olesen et al. 2007), as well as possibly from a wider a range of suitable plants coming from the south (Rochette et al. 2004). However, the growing period of determinate crops (e.g. onion, cereals) is expected to decline, contrary to indeterminate crops (e.g. carrot) (Olesen and Bindi 2002). Mediterranean-like areas are expected to show declining yield due to heat and water shortage (Olesen et al. 2011). Hereby the range, in which beneficial and degrading effects as well as the extent of these climate impact signals is still under debate. Increasing temperatures lead to a faster development of most crops (Rubino et al. 2012). This does however not necessarily imply, that – in the mean – warmer future scenarios lead to a reduction of the cultivation/production time [d], as the projected seasonal warming pattern from scenarios may differ. For example Campi et al. (2012) found a larger reduction of the production time for potato with the relatively "cooler" scenario B1, as the applied projection showed higher spring temperatures (tab.1.9) compared to scenario A2. Hereby a shorter cultivation time of field crops may influence the irrigation volume (Campi et al. 2012). Analyzing the irrigation demand for vegetables in a Mediterranean environment (scenarios A2, B1), the authors found a generally low risk of future water stress for crops which develop their cycles in spring and autumn as well as for asparagus and artichoke (summer crop cycle). However, despite shorter crop cycles, Rubino et al. (2012) found little changes of future irrigation water demand for tomato, asparagus, sugar beet and grape vine in Italy. This illustrates the counteracting effects of shorter crop cycles and increased evapotranspiration with regard to irrigation demand and due to increasing temperature. Therefore larger yields could be expected, when evapotranspiration is compensated by irrigation. Nonetheless, Olesen and Bindi (2002) compared 11 studies of agricultural crops with and without irrigation and found no clear signal by 2050. Table 1.9 shows expected trends of climate impact. The listed studies (mainly horticulture) are heavily based on plant phenology and production time, as these are closely related to temperature.

Besides detailed projections (table 1.9), rough estimates of direct climate effects have been published, e.g.

the impact of rising ozone levels on horticultural plant yield (Booker et al. 2009) and altered "competitive interactions in favor of C_3 species" due to increasing CO_2 (Bowes 1991). A further possible trend affecting plant integrity as well as grape composition/flavor is the increase of UV-B radiation, as postulated by (Schultz 2000). Further, little attendance has been paid to the consequences to assimilation caused by leaf warming beyond their thermal optimum (Sage et al. 2008). The ongoing trend of changing quality in wine production (Vršic and Vodovnik 2012) is expected to continue (Pieri et al. 2012). This is due to earlier flowering combined with strong temperature increases during the veraison-maturation period, leading to changes in aromas and phenolics (Vršic and Vodovnik 2012). However, according to Zebisch et al. (2005: 73) warming could lead to a northward shift of wine production areas in Germany, allowing the cultivation of superior varieties. Further south, these temperature effects combine with dryness, resulting in adverse conditions for viticulture in Spain (Malheiro et al. 2012). Future trends can be opposing for related effects, as both decreasing cold stress (beneficial) and decreasing winter hardiness (harmful) are described for fruit trees in Canada (Rochette et al. 2004) and horticultural woody-plants in Finland (Laapas et al. 2012). Hereby decreasing winter hardiness was attributed to a thinner snow cover (Canada) and an increasing number of thawing events (Canada and Finland). Similarly, chilling and forcing of temperate climate fruit trees may exert opposing effects in a warming world (Legave et al. 2008b,a). These findings are derived from sequential or parallel chilling-forcing models for the prediction of apple flowering dates. Hereby most studies expect an ongoing advancing of apple spring phenology. Projected changes in strawberry phenology (table 1.9) however are not expected to lead to changes in cultural techniques (Døving 2009).

While the resulting yield has been rarely estimated for horticultural crops, estimates (t ha^{-1}) are listed by table 1.10. However, absolute yields (t) depend on land use. Hereby global arable land might decrease by 0.8 to 4.4 % (2071-2099 minus 1961-1990, Europe 11 to 17 %), depending on the scenario (Zhang and Cai 2011), Olesen et al. (2007) projected an increase of suitable area for grain maize and Lorencová et al. (2013) project a "decrease of arable land" for areas in the Czech Republic. Therefore climate change effects on plant production can be somewhat different from climate change effects on human well-being. Decreasing yield of fruit trees under climate change has been estimated by Chmielewski et al. (2009) (table 1.10), finding a shift of the main sources of yield loss. While present losses are almost entirely due to blossom frost and codling moth (*Cydia pomonella*), future losses were attributed to roughly 50 % to insufficient chill as well as to direct yield losses due to water supply and shorter ripening periods. The yield of vegetables which are currently grown at supra-optimal temperatures can be expected to increase directly with rising temperature (Fink et al. 2009). However, as most vegetables are harvested according to defined weight and/or size, increasing temperature would only lead to an effective increase of total yield, if temperature would allow the cultivation of at least one additional set due to faster development/growth (Fink et al. 2009).

Greenhouse tomato yield has been projected for Avignon, Southern France, for an temperature increase of 1.0 °C (greenhouse) and 2.2 °C (ambient) for 2070-2099, as compared to 1960-1979 (Boulard et al. 2011). Hereby simulations without considering rising CO_2-concentration led to a yield reduction of 7 %, whereas considering an increase of CO_2 rendered a yield increase of 20 %. In both cases, the authors simulated a considerably increasing plant stress due to "high temperature and saturation deficit", possibly becoming problematic for fruit quality. These results were further accompanied by 30 % projected savings in greenhouse energy demand due to milder future winter. According to Fink et al. (2009), the problem of high temperatures in greenhouses can be attributed to the combination of high ambient temperature as well as high insulation and might apply only partially to higher latitudes. While the number of high ambient air temperatures is expected to increase, increasing radiation is not expected. Fink et al. (2009) conclude, that the resulting difficulties for crop cultivation in greenhouses in summer of Mediterranean climates do not apply for higher latitudes, even with increasing number of high temperatures. However, they further conclude for Germany, that energy savings due to increasing production will not lead to large changes in winter crop production in greenhouses, as the limiting factor for growth remains radiation.

Not taken into account by most studies, adaptive measures can alleviate plant stress. For instance, crop/yield losses due to heat stress were reduced in projections including adaption of sowing dates (Osborne et al. 2013; Teixeira et al. 2013). Adding to this, most studies focus on the range of projected mean temperatures of scenarios B1, A1B and A2 $<+4$ °C . However, Rötter et al. (2011) showed for *H. vulgare* L. for Finland, that warming beyond +4 °C would lead to a reduced growth duration and yield, regardless of adjusted sowing dates. Hence, climates of extreme scenarios could contain non-linear effects on plant production, yet unknown or underestimated.

Table 1.9: Expected trends of future abiotic impact of climate change on plant development or growth. W: World; Regions are abbreviated by standard code (ISO-3166-1 Alpha-2)

Factor	Species	Impact (+,-,○)[a]	Scenario	Reference
Sowing date	cereals	-	A1B	Olesen et al. (2012)
Production time	B. oleracea var. botr. L.	-	low to high[b]	Wurr et al. (2004)
Production time	B. oleracea var. it. P.	○	A2	Campi et al. (2012)
Production time	B. oleracea var. it. P.	○	B1	Campi et al. (2012)
Production time	C. scolymus	-	A2	Campi et al. (2012)
Production time	C. scolymus	○	B1	Campi et al. (2012)
Production time	Asparagus	-	A2	Campi et al. (2012)
Production time	Asparagus	-	B1	Campi et al. (2012)
Production time	S. tuberosum	-	A2	Campi et al. (2012)
Production time	S. tuberosum	-	B1	Campi et al. (2012)
Flowering date	P. serrulata 'Kwanzan'	-	A1B, A2	Chung et al. (2011)
Flowering date	P. ×yedoensis 'Yoshino'	-	A1B, A2	Chung et al. (2011)
Flowering date	M. domestica	-	A1B	Hoffmann et al. (2012)
Blossom frost risk	M. domestica	-,○	A2, B2	Eccel et al. (2009)

CHAPTER 1. INTRODUCTION 1.3. SUSCEPTIBILITY TO CLIMATIC CHANGES

Factor	Species	Impact (+,-,∘)[a]	Scenario	Reference
Blossom frost risk	M. domestica	-,∘,+	A2	Kaukoranta et al. (2010)
Blossom frost risk	M. domestica	+	—	Cannell and Smith (1986)
Blossom frost risk	M. domestica	+	B2	Chmielewski et al. (2005)
Blossom frost risk	M. domestica	∘,+	B1, A1B	Chmielewski et al. (2009)
Blossom frost risk	M. domestica	-,∘,+	A1B	Hoffmann et al. (2012)
Grow. season length	M. domestica	- (n.s.)	A1B,A2,B2	Stöckle et al. (2011)
Chill availability	—	-,∘	various	Luedeling (2012)
Chill availability	fruit/nut trees	-,∘	B1, A1B, A2	Rochette et al. (2004); Luedeling et al. (2009b)
Damage through insufficient winter chill	M. domestica	+	B1, A1B	Chmielewski et al. (2009)
Damage through early winter frost	fruit trees	-	c	Rochette et al. (2004)
Cold stress (winter)	fruit trees	-	c	Rochette et al. (2004)
Cold stress (winter)	hortic. woody-plants	-	B1, A2	Laapas et al. (2012)
Overwintering risk	hortic. woody-plants	+	B1, A2	Laapas et al. (2012)
Dormant period	—	-	A1B,A2	Vujadinović et al. (2012)
Winter hardiness	fruit trees	-	c	Rochette et al. (2004)
Bud frost risk	fruit trees	-,∘,+	c	Rochette et al. (2004)
Start of season	Fragaria ×ananassa cv. 'Senga Sengana'	-	RegClim	Døving (2009)
Duration of harvest season	Fragaria ×ananassa cv. 'Senga Sengana'	-	RegClim	Døving (2009)
Crop loss due to heavy precipitation	field crops	+	HCGS,CCGS[d]	Rosenzweig et al. (2002)
Water allocation[e]	horticultural field crops	+	A2, B1	Rubino et al. (2012)
Water allocation[e]	O. europea, B. vulgaris subsp. vulgaris	∘	A2, B1	Rubino et al. (2012)
Suitable area EU	Z. mays	+	A1F1,A2,B1,B2	Olesen et al. (2007)
Irrig. demand W	—	+	IS92a,HC3AA	Döll (2002)
Irrig. demand IT	Actinidia[f]	+	A1B	Villani et al. (2011)
Irrig. demand IT	agricultural	+	A1B	Rehana and Mujumdar (2012)

Factor	Species	Impact $(+,-,\circ)^a$	Scenario	Reference
Irrig. demand US	vegetables	+	—	USDA (1995) as quoted in Moss and De Bodisco (2002)
Irrig. demand US	orchards	+	—	USDA (1995) as quoted in Moss and De Bodisco (2002)
Irrig. demand US	agricultural	+	—	USDA (1995) as quoted in Moss and De Bodisco (2002)
O_3	—[g]	+	A2	Fuhrer (2009)
Flowering date	V. vinifera[h]	-	A1B	Pieri et al. (2012)
Budbreak to maturity time	V. vinifera[i]	-	various (8)	Bindi et al. (1996)
Bloom to maturity time	V. vinifera[i]	-	various (8)	Bindi et al. (1996)
Fruit dry weight	V. vinifera[i]	-,$\circ$,+	various (8)	Bindi et al. (1996)
Final total dry weight	V. vinifera[i]	+	various (8)	Bindi et al. (1996)
Dry matter variability	V. vinifera[i]	+	various (8)	Bindi et al. (1996)
Hailstorm damage	field crops	+	+1,+2 °C	Botzen et al. (2010)
Hailstorm damage (greenhouse)	—	+	+1,+2 °C	Botzen et al. (2010)
Heat stress	agricultural	+	A1B	Teixeira et al. (2013)
Production Suitability of Subtropics	Musa	-	A2	Van Den Bergh et al. (2012)

[a]Increase (+), decrease (-), no change ($\circ$)
[b]UK Climate Change Impacts Programme emission scenarios
[c]1 % annual increase of greehouse gas forcing, Canadian Global Coupled General Circulation Model CGCMI
[d]Hadley Centre (HC), Canadian Centre (CC) scenarios with greenhouse gas and sulfate aerosols (GS)
[e]An available amount of water was allocated to asparagus, grape vine, olive, sugar beet and tomato in Italy
[f]Not further specified
[g]Departing from present levels, that are "often sufficiently high to reduce yields of" Oryz. sat., Tritic. aest., Z. mays, Sol. tub.
[h]cv. Merlot
[i]cv. Sangiovese, cv. Cabernet S.

Table 1.10: Expected future abiotic impact of climate change on yield or production. W: World; Regions are abbreviated by standard code (ISO-3166-1 Alpha-2)

Species	Region	Δyield[a] [%]	Period/ condition	Scenario	Reference
B. vulgaris L.	UK[b]	19	2025	own[c]	Wurr et al. (1998)
B. vulgaris L.	UK[b]	32	2050	own[c]	Wurr et al. (1998)
Allium cepa L.	UK[b]	13	2025	own[c]	Wurr et al. (1998)
Allium cepa L.	UK[b]	21	2050	own[c]	Wurr et al. (1998)
D. carota L.	UK[b]	9	2025	own[c]	Wurr et al. (1998)
D. carota L.	UK[b]	13	2050	own[c]	Wurr et al. (1998)
T. aestivum L.	W	1.6	2030	A1B	Tebaldi and Lobell (2008)
Z. mays	W	-14.1	2030	A1B	Tebaldi and Lobell (2008)
H. vulgare	W	-1.8	2030	A1B	Tebaldi and Lobell (2008)
C_3 crops	W	10 to 20	550 ppm	experiment	Parry et al. (2007)
C_4 crops	W	0 to 10	550 ppm	experiment	Parry et al. (2007)
Tree biomass[b]	W	0 to 30[c]	550 ppm	experiment	Parry et al. (2007)
C_3 crops	W	5 to 20	550 ppm	simulation	Parry et al. (2007)
T. aestivum L.	various	31	700 ppm	experiment	Amthor (2001)
various[c]	various	33	700 ppm	experiment	Amthor (2001)
various herbaceous	various	38	700 ppm	experiment	Amthor (2001)
T. aestivum L.	various	10 to 28[d]	550 ppm	experiment	Long et al. (2006)
O. sativa L.	various	9 to 35[e]	550 ppm	experiment	Long et al. (2006)
G. max (L.) Merr.	various	14 to 35[f]	550 ppm	experiment	Long et al. (2006)
C_4 crops	various	0 to 27[e]	550 ppm	experiment	Long et al. (2006)
T. aestivum L.	EU	25 to 41	2020	B1, B2, A2, A1F1	Ewert et al. (2005)
T. aestivum L.	EU	37 to 101	2050	B1, B2, A2, A1F1	Ewert et al. (2005)
T. aestivum L.	EU	43 to 163	2080	B1, B2, A2, A1F1	Ewert et al. (2005)
M. domestica	DE	-7.7 to -5.7	2085	B1, A1B	Chmielewski et al. (2009)
Citrus	US	-37.1 to 526.1	+1.5, +2.5, +5.0	—	Rosenzweig et al. (1996)
S. tuberosum[j]	US	-64.0 to -1.4	+1.5, +2.5, +5.0	—	Rosenzweig et al. (1996)
S. tuberosum[k]	US	-70.8 to 3.8	+1.5, +2.5, +5.0	—	Rosenzweig et al. (1996)
Gossypium ssp. L.	IN	-1.2 to 2.9[l]	2080	A1B, A2, B2	Hebbar et al. (2013)
O. sativa	IN	-10 to >-2.5	2080	A1B, A2, B1, B2	Soora et al. (2013)
V. vinifera L.	PT	+10[m]	2085	A1B	Santos et al. (2013)

Species	Region	△yield[a] [%]	Period/ condition	Scenario	Reference
various (nut, fruit) [n]	US	<-40 to 0	2050	A1B, A2, B1	Lobell et al. (2006)

[a] △yield: difference in yield
[b] daylit cabinets
[c] derived from Houghton et al. (1992)
[d] above ground
[e] higher values mostly for younger trees
[f] 37 species, mostly agricultural
[g] range over 7 studies
[h] range over 8 studies
[i] range over 10 studies
[j] fix planting date
[l] calculated from table
[m] wine production
[n] *V. vitifera, P. dulcis, Citrus* × *sinensis* L., *J. regia, Persea americana*

Chapter 2

General objectives

The overarching objective of the present work is to assess the future climatic impact on regional horticultural production by establishing a basic frame of a climate impact modeling chain.

This objective requires investigation on processing of simulated time series including bias correction and uncertainty assessment as well as on the specific climate change impact on future horticultural production, thus comprising a broad range of substitutable methods and future projections. Therefore detailed investigations are presented by four publications. Regionalization is studied by the example of Lower Saxony (Germany) and Germany, whereas specific climate variables are exemplified by water stress, blossom frost risk in fruit tree production as well as greenhouse energy consumption.

Chapter 3

Investigations

3.1 Processing and calibration of climate input data

3.1.1 Objective

The objective is to develop a method, which allows the calibration of multiple simulated time series of different climate variables for their use as input of horticultural models.

3.1.2 Summary

A multidimensional bias correction is presented. As described in section 1.2.6, simulated climate time series tend to deviate systematically from observations. These deviations in mean, extremes and distribution skewness can decrease the applicability of raw, simulated time series for impact models drastically. Thus, correction procedures are applied to remove the bias and several approaches have been published. Nevertheless, none of them is able to maintain physical consistency between different climate variables. Accordingly, bias correction is criticized (Ehret et al. 2012). On the contrary, separate bias correction of different climate variables may lead to completely wrong results of impact models. For instance, separate correction of precipitation and global radiation decreased the fraction of diffuse radiation of hours with precipitation ($\geq$0.1 mm, Hoffmann and Rath 2011). Hereby, consistent bias correction has been identified as one of six end user's needs regarding the downscaling of climate data (Maraun et al. 2010). As most plant models require more than one climate variable as input, a consistent bias correction method is required.
The presented bias correction method optimizes bias and n-dimensional probability for n climate variables. Hereby single climate variable bias correction is applied and absolute bias and probability are weighted for optimization. While this approach can be based on any bias correction method, the present work employs distribution based bias correction (Ines and Hansen 2006; Piani et al. 2010) with hourly global radiation and precipitation. The latter were chosen in order to show the strong relationship of the climate variables.
Depending on bias and consistency prior to correction as well as on the purpose of correction, optimization may not exploit the full range of possibilities. This is due to weighting of probability and bias. Therefore a parameter K was introduced, allowing to adjust the weights for bias and probability. For the time series investigated, bias was reduced similarly to separate quantile mapping, while consistency as estimated from the bivariate empirical cumulative distribution was improved.

3.1.3 Publication: Meteorologically consistent bias correction of climate time series for agricultural models — *Theoretical and applied climatology*

Authors:	Holger Hoffmann[1] and Thomas Rath[2]
Journal:	Theoretical and Applied Climatology
Volume:	110
Page:	129-141
DOI:	10.1007/s00704-012-0618-x
ISSN:	0177-798X (print version), 1434-4483 (electronic version)
Publisher:	Springer
Address:	Vienna, Austria
Date of Submission:	28 June 2011
Date of Acceptance:	22 February 2012
Date of Publication:	14 March 2012
Current Status:	Published

Contribution of authors:

1: Data acquisition and processing, calculations, evaluation, manuscript development
   and processing

2: Review

This publication can be accessed via `http://link.springer.com/article/10.1007/s00704-012-0618-x` and is indexed by 'scopus','web of knowledge','google scholar' among other search engines.

3.2 Future water stress risk for *Lactuca sativa* L. var. *capitata*

3.2.1 Objective

The objective is to assess the influence of decreasing water availability on the growth of *Lactuca sativa* L. var. *capitata*. For this purpose, a growth model was developed as described in the following. Future climate impact due to changes in precipitation are discussed in section 4.2.

3.2.2 Summary

Leafy crops are highly susceptible to changes in the available water regime. Hereby future changes in total precipitation (mm) and precipitation intensity (mm d^{-1}) as well as in the precipitation pattern of the number of consecutive wet or dry days are expected. Decreasing total summer precipitation along with longer dry periods followed by heavy precipitation might pose problems to horticultural open field crops in the long run. In order to cope with these changes, detailed knowledge about plant responses in cropping systems is essential. Knowing the extent of plant stress at specific growth stages is a starting point for developing suitable adaptation strategies. Hereby the effects of mean soil water tension (SWT) for a given period must be separated from effects of SWT falling below a given threshold for a given time and/or frequency. As the effects of resulting water stress may be specific to species and growth stage, these thresholds and corresponding transition zones are largely unknown. Hereby severe plant stress due to water scarcity is and must be avoided through irrigation. In order to enable the projection of future precipitation change effects, a first empirical model accounting for the cultivation time and yield response to mean precipitation was developed.

The following publication describes the experimental set-up and model development for lettuce (*Lactuca sativa* L. var. *capitata*). Growth was simulated with a parametrized double gompertz-function of soil temperature and global radiation and water stress. Yield decreases due to water scarcity were modeled as deviation from optimum growth (control), introducing a stress factor. Hereby the stress factor is a function of soil moisture which was simulated from evapotranspiration and precipitation/irrigation. This approach accounts for additive and compensating feedback-effects of soil moisture→ growth → soil moisture. Future projections must however be supplied with initial soil moisture. Furthermore, the model yet lacks of differentiation of drought situations regarding growth stage and duration which might result in different results as plant adaptation (compounds, root-shoot ratio, fine roots etc.) takes place. In spite of these limits of the area of application, estimates of future precipitation impacts seem feasible.

3.2.3 Publication: Dynamic Modelling of Water Stress for *Lactuca sativa* L. var. *capitata* — *Acta Horticulturae*

Authors:	Charlotte Duncker[1], Andreas Fricke[2], Thomas Rath[3] and Holger Hoffmann[4]
Journal:	Acta horticulturae
Volume:	—
Page:	—
DOI:	—
ISSN:	0567-7572
Publisher:	International Society for Horticultural Science (ISHS)
Address:	Leuven, Belgium
Date of Submission:	25.05.2012
Date of Acceptance:	09.07.2013
Date of Publication:	—
Current Status:	Accepted

Contribution of authors:

1: Data acquisition, experiments, model development and parametrization, simulations, evaluation, manuscript development and processing

2: Expert knowledge, experiment design

3: Review

4: Model development, manuscript development, review

Once published, this publication will be accessible via `http://www.actahort.org/` and will be indexed by 'scopus' and 'google scholar' among other search engines. A personal preprint version is given in the following.

Dynamic Modelling of Water Stress for *Lactuca sativa* L. var. *capitata*

Charlotte Duncker[1]*, Andreas Fricke[2], Thomas Rath[1] and Holger Hoffmann[1]

Institute for Biological Production Systems, [1] Biosystems and Horticultural Engineering Section, [2] Institute for Vegetable Systems Modelling, Leibniz University Hannover, Herrenhäuser Str.2, 30419 Hannover, Germany.

* E-mail: charlotte.duncker@gmx.de

Abstract

Changes in precipitation patterns and water availability are expected to continuously increase the importance of efficient and reliable irrigation management in intensive plant production. Therefore accurate prediction of plant growth for deficient soil moisture conditions is essential. A dynamic model was developed in order to investigate the impact of different water regimes on the growth of lettuce. For this *Lactuca sativa* L. var. *capitata* was grown in a semi-open greenhouse at two irrigation levels (30 % and 80 % of potential evapotranspiration) and growth parameters were derived by relating fresh matter to soil and climatic conditions. The developed model is built modularly and growth under optimal conditions is simulated by a two-dimensional dynamic model based on soil temperature and global radiation. Water deficiency – in turn affecting growth – is taken into account by subsequent computing of soil water status as influenced by growth and evapotranspiration. In addition to the optimization of irrigation strategies the model is further designed to be employed in climatic impact projections.

Keywords: growth model, drought stress, irrigation, evapotranspiration, lettuce

Introduction and Objectives

Water shortage leads to one of the most common forms of stress in plants (Molina-Montenegro et al., 2011). Hence, water management is essential in plant production (Capra et al., 2008) and its importance increases with the climatic change due to ongoing scarcity of fresh water (Leenhardt et al., 1998). Especially for intensive plant production efficient irrigation methods become necessary (Capra et al., 2008). Leafy crops as lettuce are sensitive to drought stress (Coelho et al., 2005). As the effect of latent water deficiency on yield with regard to the interdependencies of time and period of deficiency, plant size and age, soil moisture tension and climate among others remains partially unknown, models might be used to select adequate irrigation strategies. Considering the complexity of plant-soil-water interactions, dynamical-empirical models are suitable tools to provide an insight to these relations through continuous calculation of compartment conditions of the observed system. At this, approaches to simulate lettuce at the cellular (Ioslovich et al.,

2002), plant or canopy (Leal-Enriquez and Bonilla-Estrada, 2010) level have been developed for varying situations as protected cultivation in greenhouses (Coelho et al., 2005), including cultivation in hydroponics or pots (Xu et al., 2004). Additionally, growth and physiology of lettuce grown in soil (greenhouse or open soil) has been monitored (Capra et al., 2008). Also soil moisture models as for example HYDRUS 1D (Jiménez-Martínez et al., 2009) are available. Nevertheless, complex models accounting for soil water conditions and therefore growth of lettuce at the whole plant level under water shortage have not been described so far. This could be due to the sophisticated task of synchronizing plants and soil measurements with the corresponding physics. Basic soil moisture simulation methods as the "Tipping Bucket" approach could be used (Conolly, 1998). Hence, the objectives of this work are to develop a basic plant-soil-model for lettuce and to simulate growth reduction under drought conditions. In order to estimate the effects of latent water deficiency at an early growth stage of lettuce, measurements were used to model plant growth for open soil conditions.

Materials and Methods

Open Soil Measurements

In order to determine the growth of Lettuce under water deficit conditions, an open soil experiment was conducted inside a greenhouse which served as a rain out shelter. Being completely open, the greenhouse allowed free passage of the wind while excluding precipitation. Seedlings of *Lactuca sativa* L. var *capitata*, cv. 'Centore' were planted 32 days after germination in sandy soil (65 % sand, 28 % silt, 7 % clay, 180 kg N ha^{-1}, permanent wilting point 6 Vol.-%, field capacity 22 Vol.-%). Mean precipitation conditions where simulated by irrigating plants every other day in treatments of replacing 30 or 80 % of water lost due to evapotranspiration. The latter was calculated using the equation from Penman-Monteith (Allen et al. 1998). For this and in order to derive growth parameters the following climate variables were measured continuously: radiation [W] (solarimeter), wind speed [m s^{-1}] (anemometer), air- and soil-temperature [°C] (NTC-sensors), relative air humidity [% rh] (psychrometer) and soil moisture [Vol.-%] (TDR-sensors). Soil moisture measurements were checked with readings from FDR-sensors (ML2-Theta-Probe). Treatments were conducted in four replications and from each replication five plants were harvested every seventh day and fresh and dry weight as well as diameter per plant were determined. Multiple contrast tests (MCP, Tukey comparison) were used to test results.

Growth Model and Parameter Estimation

A dynamic model was implemented to simulate lettuce growth during changing water availability. Plant growth, soil moisture and water uptake are therefore linked to each other (Fig. 1). Soil moisture is calculated through continuous computation of the daily balance between evapotranspiration and irrigation, considering further the percolation in two soil layers z from 0 - 15 and from 15 - 30 cm depth (please see Tab. 1 for abbreviations):

$$C_{n,z} = \begin{cases} B_n - ETa_n - P_{n,z} & if \quad z = 1 \\ -P_{n,z} + P_{n,z-1} & else \end{cases} \quad (1)$$

$$CS_{i,z} = \sum_{n=1}^{i} C_{n,z} \cdot u \quad (2)$$

$$CS_i = \frac{CS_{i,1} + CS_{i,2}}{2} \cdot \frac{2}{3} \quad (3)$$

CS_i is calculated in Vol.-% considering a layer depth of 150 mm. The percolation rate is calculated for one day as follows:

$$P_{n,z} = \begin{cases} 0 & if \quad CS_{i,z} < FC \\ CS_{n,z} - FC & else \end{cases} \quad (4)$$

ETa_i is a function of ETc_i and a stress factor function of CS_i with resulted in ks (Allen et al. 1998 and Fig. 3) for non optimal water conditions:

$$ETa_i = ETc_i \cdot ks \text{ with } ks = f_A(CS_i) \quad (5)$$

ETc was calculated depending on a specific plant factor k_c (Allen et al. 1998) which is a function of the plant diameter (Eq. 6).

$$k_c = \begin{cases} 0.5 & if \quad D_i < 20 \\ 0.8 & if \quad 20 \leq D_i < 28 \\ 1.2 & else \end{cases} \quad (6)$$

where the plant diameter itself was derived with a regression equation from measured fresh weight as follows (Fig. 2):

$$D_i = 17.1518 + 1.1889 \cdot W_i^{0.4898} \quad (7)$$

In order to reproduce plant growth as influenced by soil moisture, a bivariate growth function (Salomez and Hofmann, 2008) for growth under optimal conditions (ETc 80 %) was parametrized (Eq. 8). Growth reduction due to changes in soil moisture were accounted for by multiplicating with a stress factor $f_M(CS_i)$, which was obtained by relating final fresh weight to the mean soil moisture during the experiment (Fig. 3):

$$W_i = f_M(CS_i) \cdot \left(a \cdot e^{b \cdot e^{c \cdot TS_i}} + d \cdot e^{g \cdot e^{h \cdot RS_i}}\right) \text{ with} \tag{8}$$

$$TS_i = \sum_{n=1}^{i} T_n \cdot u$$

$$RS_i = \sum_{n=1}^{i} R_n \cdot u$$

$$f_M(CS_i) = \text{General plant growth stress factor (see Fig.3)}$$

Parametrization of a, b, c, d, g and h of equation 8 was done by using the Simplex Algorithm (Nelder and Mead, 1965). The model was implemented in Model Maker v3.0.3 (Cherwell Scientific Publishing, Germany, Frankfurt).

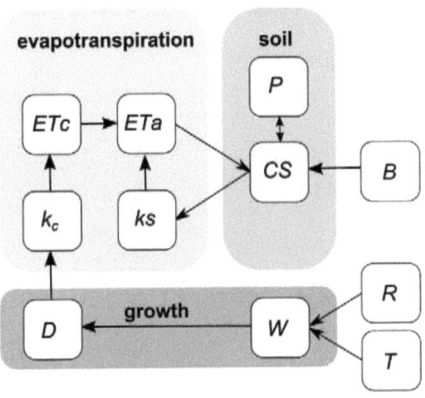

Figure 1: Sketch of developed water stress model for *Lactuca sativa*. For abbreviations see Tab. 1.

Simulation of lettuce growth

In order to estimate the influence of latent water stress during an early growth stage (0 to 35 days after planting) on the final yield of lettuce, plant growth was simulated depending on daily global radiation and soil temperature for well watered (80 % ETc) and water deficient (30 % ETc) conditions. Hereby measured plant growth was reproduced with the help of measured climatic data and extrapolated by continuing the simulation until reaching marketable yield (>250 g fresh weight). For this the following mean climatic data were used: Global radiation (mean 11.6 MJ m^{-2} d^{-1}), soil temperature (mean 12.86 °C), air temperature (mean 12.59 °C), relative air humidity (mean 53.56 % r.H.) and wind speed (canopy, mean 0.51 m s^{-1}). W_i, ETa_n and CS_i were solved iteratively with the integration method Runge-Kutta (Abramowitz, Stegun 1972).

Results

Water supply led to considerable differences in soil and plant conditions, which is reflected in the measured soil moisture (Fig. 4). Whereas soil moisture of the control treatment did not decrease on average, dry treatment soil moisture decreased continuously to 12.7 Vol.-% on day 36 after planting. Consistently plants of the control treatment exhibited increased fresh and dry weight as well as plant diameter compared to plants irrigated at 30 % ETc (Fig. 5). The pattern of measured soil moisture [Vol.-%] was reproduced in the main, with better reproduction of the deficient irrigation treatment (RMSE = 0.52 and RMSE = 1.09 respectively; Fig.4). In both treatments simulated soil moisture responded slower to changes of the soil water status than measured. Furthermore, measured fresh weight per plant [g] was reproduced by simulation (RMSE = 11.44 and RMSE = 8.89 for control and deficient irrigation treatment respectively). Simulation of fresh weight beyond the measured time span (Fig. 5) resulted in 50 g less fresh weight per plant of the dry variant at harvest time.

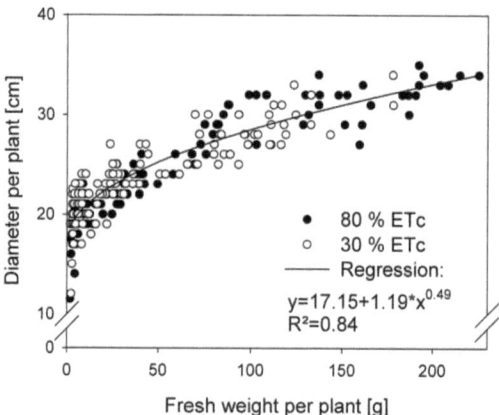

Figure 2: Relationship between fresh weight and plant diameter.

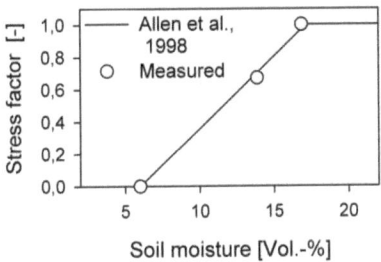

Figure 3: Stress factor as a function of soil moisture. $f_A(CS_i)$: Allen et al., $f_M(CS_i)$: measured.

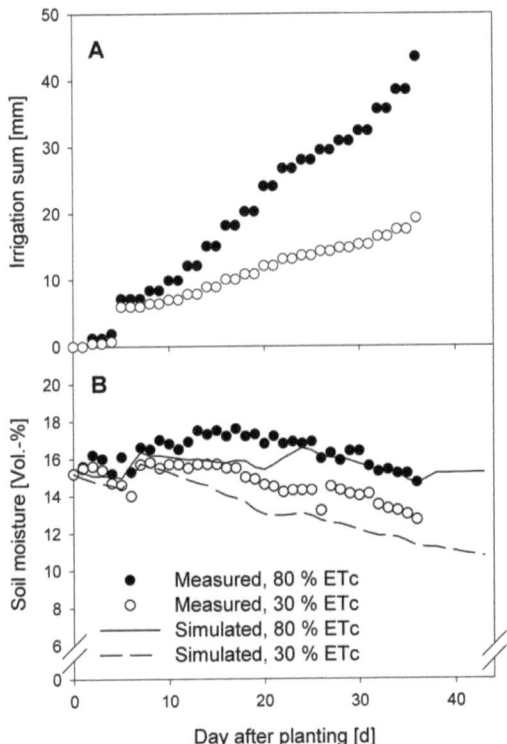

Figure 4: Irrigation management of the treatments (A) and development of measured (TDR) and simulated soil moisture (B, 0-30 cm).

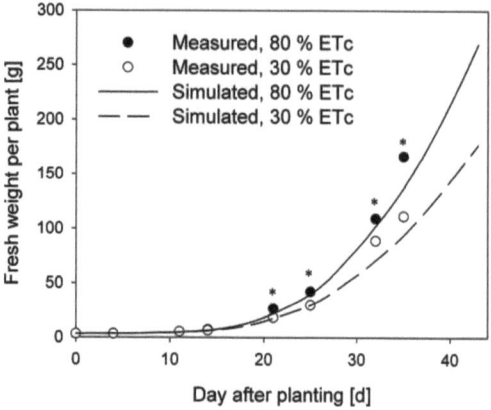

Figure 5: Influence of the irrigation treatment on measured plant growth and simulated plant growth. Significant differences are indicated with * (MCP, α = 0.05). Parameters a = $2.4e^{-14}$, b = -21.3, c = 10.0, d = 1679.1, g = -9.5, h = $-3.1e^{-3}$.

Discussion and Conclusion

Soil moisture of the deficient irrigation treatment led to water stress while plants of the control treatment were not affected. The last harvest of this experiment revealed a decrease in yield of about 20 % at a mean soil moisture of 14.7 Vol.-%. These findings are in agreement with Coelho et al. (2005) who found lettuce to be wilting after deficient irrigation and mean soil moisture of 13.9 Vol.-%. The induced water stress led to reduced fresh and dry weight, similar as observed by Arkhipova et al. (2007) and Ruiz-Lozano et al. (2011) in their experiments. As in the present work, a decrease of dry weight of approximately 20 % was determined under water stress conditions. At this, fresh weight was reproduced during the early growth stage but was underestimated by the simulation beyond 30 days after planting. Hereby the parameters (a-g) were fitted to a growth function derived from plants grown at optimal water supply (80 % ETc data), excluding CS_i from fitting. In the following, for both treatments small deviations in soil moisture added up to the apparent differences at the end of the experiment. Nevertheless simulated soil moisture exhibited the same pattern as observed. Hence, despite underestimation, the drying process of the soil was reproduced in the main, resembling findings for drying soil under lettuce plants from Jiménez-Martinez et al. (2009). However, as model development was based on a single experimental data set, a generalization of the present findings is not possible without verification through future experiments. Finally, for this purpose varying climatic situations have to be accounted for.

References

Abramowitz, M.; Stegun, I.A. 1972. Handbook of Mathematical Functions with Formulas, Graphs, and Mathematical Tables, 9th printing. 896-897

Allen, R. G.; Pereira, L. S.; Raes, D.; Smith, M. 1998. Crop evapotranspiration - Guidelines for computing crop water requirements - FAO irrigation and drainage paper 56. FAO - Food and Agricultur Organization of the United Nations. www.fao.org

Arkhipova, T. N.; Prinsen, E.; Veselov, S. U.; Martinenko, E. V.; Melentiev, A. I.; Kudoyarova, G. R. 2007. Cytokinin producing bacteria enhance plant growth in drying soil. Plant Soil 292. 305-315

Capra, A.; Consoli, S.; Russo, A.; Scicolone, B. 2008. Integrated Agro-Econcomic Approach to Deficit Irrigation on Lettuce Crops in Sicily (Italy). J Irrig Drain E 134. 437-445.

Coelho, A. F. S.; Gomes, E. P.; Sousa, A. d. P. ; Glória, M. B. A. 2005. Effect of irrigation level on yield and bioactive amine content of American lettuce. J Sci Food Agr 85. 1026-1032

Conolly, R.D. 1998. Modelling effects of soil structure on the water balance of soil-crop systems: a review. Soil Till Res 48. 1-19

Gallardo, M.; Jackson, L.E.; Thompson, R.B. 1996. Shoot and root physiological responses to localized zones of soil moisture in cultivated and wild lettuce (Lactica spp.), Plant Cell Environ 19. 1169-1178

Ioslovich, I.; Seginer, I.; Baskin, A. 2002. Fitting the Nicolet Lettuce Growth Model to Plant-spacing Experimental Data. Biosystems Engineering 83. 361-371

Jiménez-Martínez, J.; Skaggs, T.H; van Genuchten, M.T.; Candela, L. 2009. A root zone modelling approach to estimating groundwater recharge from irrigated areas. J Hydrol 367. 138-149

Leal-Enríquez, E.; Bonilla-Estrada, M. 2010. Modelling the greenhouse lettuce crop by means of the daily interaction of two independent models. IEEE 49. Conference on Decision and Control. 4667-4672

Leenhardt, D.; Lafolie, F.; Bruckler, L. 1998. Evaluating irrrigation strategies for lettuce by simulation: 1. Water flow simulations. Eur J Agron 8. 249-265

Molina-Montenegro, M. A.; Zurita-Silva, A.; Oses, R. 2011. Effect of water availability on physiological performance and lettuce crop yield (Lactuca sativa). Ciencia e Investigación Agraria 38. 65-74

Nelder, J. A.; Mead, R. 1965. A Simplex Method for Function Minimization. Comput J 7. 308-313

Ruíz-Lozano, J. M.; Perálvarez, M. d. C.; Aroca, R.; Azón, R. 2011. The application of a treated sugar beet waste residue to soil modifies the responses of mycorrhizal and non mycorrhizal lettuce plants to drought stress. Plant Soil 346. 153-166

Salomez, J.; Hofman, G. 2007. A Soil Temperature/Short-Wave Radiation Growth Model for Butterhead Lettuce Under Protected Cultivation in Flanders. J Plant Nutr 30. 397-410

Xu, G.; Levkovitch, I; Soriano, S.; Wallach, R.; Silber, A. 2004. Integrated effect of irrigation frequency and phosphorus level on lettuce: P uptake, root growth and yield. Plant Soil 263. 297-309

Appendix

Table 1: Abbreviations

Variable	Description	Unit
a, b, c, d, g, h	regression parameters	—
B_n	irrigation at day n	mm
$C_{n,z}$	change of soil water content in layer z at day n	mm d^{-1}
$CS_{i,z}$	soil water content in layer z at day i	mm
CS_i	soil water content at day i	Vol.-%
D_i	plant diameter	cm
ETa_i	actual evapotranspiration at day n	mm
ETc_i	crop evapotranspiration at day n	mm
FC	field capacity per layer, here $FC = 33$	mm
$f_M(CS_i)$	general plant growth stress factor	—
k_c	plant specific factor	—
$ks(f_A(CS_i))$	evapotranspiration stress factor	—
n, i	day, days	d
$P_{n,z}$	percolation rate in layer z at day n	mm d^{-1}
R_n	daily average short wave radiation at day n	MJ m^{-2}
RS_i	accumulated short wave radiation	MJ m^{-2} d
T_n	daily average soil temperature at day n	°C
TS_i	accumulated soil temperature	°C d
u	integration time, here $u = 1$	d
W_i	fresh weight per plant	g

3.3 Future bloom and blossom frost risk for *Malus domestica*

3.3.1 Objective

The objective is to assess the climate impact of temperature on future bloom and blossom frost risk for regional apple production. This objective comprises the assessment of changes in the fulfillment of the chilling requirement. A secondary objective is to put the climate impact signal in relation to the range of model realizations, thus giving the magnitude of projection uncertainty.

3.3.2 Summary

Blossom frost directly affects fruit yield. How climate change will affect future blossom frost remains unknown, as both last spring freeze as well as flowering phenology advance to earlier dates in the year. Hereby flowering depends on the influence and interaction of chilling requirement and the forcing phase (see section 1.3.5). The few estimates of future blossom frost risk that have been published diverge regionally, being mostly based on single climate realizations, lacking further of statistical and/or uncertainty analysis. Additionally, transferring parametrized models from literature to different regions holds unknown risks. Thus a robust approach is presented in order to estimate regional future blossom frost risk. This includes model improvement, as simple thermal time ("heat sum") models ignore chilling requirement, whereas complex chilling-forcing models ignore photoperiodism. Since photoperiodism of apple has been stated (Körner 2006: 63) and improvement of a simple thermal model was found when including length of day (Blümel and Chmielewski 2012), a natural step would be to modify chilling-forcing models in order to account for the length of day.

Future blossom frost risk is estimated for the SRES emission scenario A1B. An ensemble approach with 13 climate realizations as well as 7 impact models (including 2 chilling-forcing models with modification for length of day) for calculation of flowering dates was chosen for this purpose. Impact models were calibrated for each grid point with observations. Projected single grid point time series were analyzed and the joint signal for the state of Lower Saxony (Germany) was segmented by variance decomposition in order to obtain information on the relative uncertainties of internal variability, climate and impact models. Contrary to literature, a mean decrease of blossom frost risk beyond 2045 was projected by single time series. This is partially due to decreasing winter chill availability. However, regarding signal stability, a large fraction of variance was caused by internal variability – as expected for highly resolved regional projections. Variance of climate realizations was larger than of impact models for the most part of the 21st century, both leveling off at the end of the century. Hereby internal and model variability of temperature, bloom and blossom frost risk exhibited an optimum projection horizon for the range 2078 to 2087. However, the variability in blossom frost risk exceeded the projected signal. Thus it can only be concluded, that future blossom frost risk is unlikely to increase. By showing the significant time horizon, the limits and meaningful range of projection are exemplified.

3.3.3 Publication: Future bloom and blossom frost risk for *Malus domestica* considering climate model and impact model uncertainties — *PLoS ONE*

Authors:	Holger Hoffmann[1] and Thomas Rath[2]
Journal:	PLoS One
Volume:	8(10): e75033
Page:	1-13
DOI:	10.1371/journal.pone.0075033
ISSN:	1932-6203 (electronic version)
Publisher:	Public Library of Science
Address:	San Francisco, US / Cambridge, UK
Date of Submission:	23.04.2013
Date of Acceptance:	09.08.2013
Date of Publication:	08.10.2013
Current Status:	Published

Contribution of authors:

1: Data acquisition and processing, calculations, evaluation, manuscript development and processing

2: Review

This open access publication is accessible via `http://www.plosone.org/` and is indexed by 'scopus','web of knowledge' and 'google scholar' among other search engines.

Future Bloom and Blossom Frost Risk for *Malus domestica* Considering Climate Model and Impact Model Uncertainties

Holger Hoffmann*¤, Thomas Rath

Biosystems Engineering, Institute for Biological Production Systems, Leibniz Universität Hannover, Hannover, Germany

Abstract

The future bloom and risk of blossom frosts for *Malus domestica* were projected using regional climate realizations and phenological (=impact) models. As climate impact projections are susceptible to uncertainties of climate and impact models and model concatenation, the significant horizon of the climate impact signal was analyzed by applying 7 impact models, including two new developments, on 13 climate realizations of the IPCC emission scenario A1B. Advancement of phenophases and a decrease in blossom frost risk for Lower Saxony (Germany) for early and late ripeners was determined by six out of seven phenological models. Single model/single grid point time series of bloom showed significant trends by 2021–2050 compared to 1971–2000, whereas the joint signal of all climate and impact models did not stabilize until 2043. Regarding blossom frost risk, joint projection variability exceeded the projected signal. Thus, blossom frost risk cannot be stated to be lower by the end of the 21st century despite a negative trend. As a consequence it is however unlikely to increase. Uncertainty of temperature, blooming date and blossom frost projection reached a minimum at 2078–2087. The projected phenophases advanced by 5.5 d K^{-1}, showing partial compensation of delayed fulfillment of the winter chill requirement and faster completion of the following forcing phase in spring. Finally, phenological model performance was improved by considering the length of day.

Citation: Hoffmann H, Rath T (2013) Future Bloom and Blossom Frost Risk for *Malus domestica* Considering Climate Model and Impact Model Uncertainties. PLoS ONE 8(10): e75033. doi:10.1371/journal.pone.0075033

Editor: Vanesa Magar, Plymouth University, United Kingdom

Received April 23, 2013; **Accepted** August 8, 2013; **Published** October 8, 2013

Copyright: © 2013 Hoffmann, Rath. This is an open-access article distributed under the terms of the Creative Commons Attribution License, which permits unrestricted use, distribution, and reproduction in any medium, provided the original author and source are credited.

Funding: The study was supported by the Ministry for Science and Culture of Lower Saxony within the network KLIFF - climate impact and adaptation research in Lower Saxony. The funders had no role in study design, data collection and analysis, decision to publish, or preparation of the manuscript.

Competing Interests: The authors have declared that no competing interests exist.

* E-mail: hhoffmann@uni-bonn.de

¤ Current address: Institute of Crop Science and Resource Conservation (INRES), University of Bonn, Bonn, Germany

Introduction

Apple production and its economic efficiency are clearly influenced by blossom frosts [1]. In addition, global warming could increase the risk due to greater changes in the date of flowering than in the last spring freeze or increasing variability in both. A generally higher risk of frost after bud burst for warmer winters was further stated as due to faster completion of the chilling requirement [2]. Past observations of late frosts and blossom frosts around the world have indicated a decreasing [3,4] up to increasing risk [4–8] for fruit trees. However, findings cannot be generalized as they vary regionally. For instance, observed damages due to late frost increased in Northern Japan while other regions of Japan exhibited different tendencies [4]. An analysis of meteorological and phenological records of the Rhineland fruit-growing region in the West of Germany revealed, that risk of apple yield loss due to frosts in April remained unchanged during the period 1958 to 2007 [9–11]. This is consistent with studies showing an advance during the past of about 2.2 d/decade for both the last spring freeze (≤0°C, Central Europe, 1951–1997) [12] and for apple flowering (BBCH 60 [13], Germany, 1961–2000) [14].

Regardless of its development during the past, future blossom frost risk development remains uncertain as published estimates diverge (Table 1). Discrepancies are mainly due to differences in selected regions and varieties, as well as to the fact, that blossom frost risk computation requires estimates for flowering dates in addition to consistent climate time series which reproduce temperature thresholds (e.g. 0°C) accurately. For this purpose climate model temperature time series are used as input for empirical phenological models accounting for chilling and/or forcing phases in winter and spring respectively [15]. While most climate scenarios describe an enhanced warming beyond 2040 [16], the following risk estimates are given. For the apple cultivar *Golden delicious* a "decreasing trend ... of little significance" was found (Trentino, Italy), concluding that blossom frost risk "will not differ greatly from its present level" [17]. Similarly, for Finland the risk is expected to generally "stay at the current level or to decrease" for the period 2011–2040 compared to 1971–2000, excepting the southern inland which exhibits increases [18]. Increases in frost damage to apple blossom (*Malus pumila* Mill. cv. Cox's Orange Pippin) were estimated for Britain [19] and an increase in the frequency of apple blossom frost damage was projected for Saxony (East Germany) by applying a simple thermal model to predict flowering, beginning on each 1 January [20]. Using the same approach, no increase in the mean apple blossom frost risk for Lower Saxony (Saxony and Lower Saxony are non-

Table 1. Published projections of future apple blossom frost risk.

Region	Increase (+) Decrease (−) No change (°)	Model	Statistics on time series[a]	Ref.
Trentino, Italy	−, °[b]	Modified Utah	yes	[17]
Finland	−, °,+[b]	Thermal Time	no	[18]
Britain	+	Thermal Time-Chilling	no	[19]
Saxony, Germany	+	Thermal Time	no	[20]
Lower Saxony, Germany	−, °,+[b]	Thermal Time	yes	[21]

[a]Tests on blossom frost risk.
[b]depending on subregion.
doi:10.1371/journal.pone.0075033.t001

adjacent states) was found [21], despite temporarily/regionally increasing blossom frost risk.

These differences in estimates can be attributed to two deficits:

1) The modeling properties of the mentioned model [20,21] are very limited for climate impact studies, as it solely calculates the onset of a phenophase based on accumulation of a heat requirement (forcing), hence assuming that dormancy has already been satisfied by a fixed starting date (see [22] for more details). Since future fulfillment of dormancy cannot be guaranteed, models including chilling phases seem to be more suitable for future climate impact simulations [23]. With their help, a possible impact of climate change on the fulfillment of dormancy [6] can be assessed. However, most of these models rely only on air temperature, ignoring possible influences of other climatic variables. Nevertheless improvement was found after including light conditions in the form of day length [24,25], despite ongoing discussions about the influence of light conditions on tree phenological phases [26].

2) Published estimates of future blossom frost risk (Table 1) are based on single climate realizations and out of five studies, only two presented statistics for future blossom frost risk [17,21]. However, assessing climate impact on the basis of models involves error concatenation resulting from the following chain of information. The future climatic impact is studied with the help of simulated climate time series, generated by global circulation models (GCM) and regionalized or downscaled by regional climate models (RCM). For this purpose these climate models are forced with greenhouse gas emissions scenarios of an evolving world (IPCC scenarios, SRES emission scenarios, [16,27]). In order to estimate climate projection uncertainty, ensembles of GCM-RCM combinations or several realizations of one GCM-RCM combination (runs) are usually produced. These climate time series are used after down-scaling to drive impact models in order to assess the climatic impact in such different fields as coastal protection, water management, environmental research, food supply, urban planning and land use. Since models cannot reproduce every environmental aspect in real accuracy and resolution, systematic deviations of simulated and observed climate time series as well as of simulated and observed climatic impact have to be taken into account. Depending on model sensitivity and question at hand, these biases can be removed by bias correction (e.g. 1-dimensional [28]; 2-dimensional [29]). Hence the chain of information for climate impact is: Scenario - emission - GCM - RCM - climate run - (bias correction) - impact model. Further chain members (e.g. prevention, adaptation strategies) or influences (e.g. feedbacks, interpolation, statistics) are possible. Since each member of this chain exists in different versions, numerous computations have to be conducted in order to cover the whole set of information available. Therefore most impact studies focus on "likely" scenarios [30], often not considering the full range of possibilities. This leads to the effect of possibly biased but significant trends of single or similar time series.

Taking these deficits into account, the objective of this work is to present a robust estimate of future blossom frost risk, taking the climate-model-impact-model uncertainty into account, including two new developed extensions of one sequential and one parallel chilling-forcing model considering light conditions.

Methods

General Procedure and Regional Focus

Thirteen simulated time series of air temperature from varying regional climate models were used to drive seven phenological models for the projection of apple bloom in Lower Saxony, Germany, whereas blossom frost risk was obtained by evaluating the temperature following bloom. Changes of these variables over time and compared to a reference period are referred to as "signal" in the following. The behavior of signal and variance across climate and impact models was analyzed subsequently, extracting the fractional uncertainty (inverse of signal-to-noise ratio). From this the meaningful horizon of projection was obtained, being basically the year at which the investigated signal exceeds the variation of the signal. This climatological approach [31,32] originally divides time series into their internal variability, scenario and model uncertainty. Advancing this approach beyond climatology, the present work estimates the extension of uncertainty from the climate signal to the climatic impact by dividing time series into their internal variability, climate model and impact model uncertainty of one scenario.

In order to project apple bloom, phenological models were calibrated with measurements of daily air temperature and observations of phenophases. Subsequent projection of future apple bloom was carried out with bias-corrected climate projections from physical-dynamical regional climate models (Table 2). Calibrated models were validated for accuracy in prediction of bloom by cross-validation as well as testing for different locations. Blossom frost risk estimates were validated first by calculating the accuracy of the phenological model (comparing measured blossom frost risk with blossom frost risk simulated with measured temperature) and secondly through calculating the influence of

Table 2. Overview of employed data.

Data	Specification	Climate model runs	Resolution (spatial, temporal)	Period	Ref.
observed flowering (DOY)	early ripeners, BBCH 60		0.116°, d	1991–2012	a
	early ripeners, BBCH 65		0.116°, d	1991–2012	a
	late ripeners, BBCH 60		0.116°, d	1991–2012	a
	late ripeners, BBCH 65		0.116°, d	1991–2012	a
measured T (°C)[b]	115 stations		0.126°, d	variable	c
simulated T (°C)[b]		1. EH5-REMO5.7, C20 1/A1B 1[d]	0.088°, h	1951–2100	[58]
		2. EH5-REMO5.8, C20 1/A1B 2[e]	0.088°, h	1961–2100	[59]
		3. EH5-REMO2008, C20 3/A1B 3[f]	0.088°, h	1950–2100	f
		4. EH5-CLM2.4.11 D2 C20 1/A1B 1	0.165°, 3 h	1961–2100	[60]
		5. EH5-CLM2.4.11 D2 C20 2/A1B 2	0.165°, 3 h	1961–2100	[61]
		6. C4IRCA3_A1B_HadCM3Q16	0.223°, d	1951–2099	[62]
		7. CNRM-RM5.1_SCN_ARPEGE	0.232°, d	1951–2100	[62]
		8. DMI-HIRHAM5_BCM_A1B	0.223°, d	1961–2099	[62]
		9. DMI-HIRHAM5_A1B_ARPEGE	0.223°, d	1951–2100	[62]
		10. DMI-HIRHAM5_A1B_ECHAM5	0.223°, d	1951–2099	[62]
		11. ICTP-REGCM3_A1B_ECHAM5_r3	0.232°, d	1951–2100	[62]
		12. KNMI-RACMO2_A1B_ECHAM5_r3	0.223°, d	1951–2100	[62]
		13. MPI-M-REMO_SCN_ECHAM5	0.223°, d	1951–2100	[62]

[a]German Meteorological Service. Phenological observation program. URL: http://www.dwd.de (April 20, 2013).
[b]air temperature at 2 m elevation.
[c]German Meteorological Service. Station network. URL: http://www.dwd.de (April 20, 2013).
[d]"UBA"-Run, experiments 6215/6221.
[e]"BFG"-Run, experiments 29001/29002.
[f]experiments 1518/1518, Max Planck Institute for Meteorology, Hamburg, Germany.
doi:10.1371/journal.pone.0075033.t002

the time series on blossom frost risk projection accuracy (comparing simulated blossom frost risk from measured temperature with that from simulated temperature).

Climatic Data and Models

Data sources. Measured as well as simulated air temperature time series for Lower Saxony, Germany, (Table 2, Figure 1) were processed and applied as follows. Simulated temperature of regional climate model projections of the IPCC-emission A1B [27] was obtained from the Max Planck Institute for Meteorology, Hamburg, Germany, (in the following climate runs 1–5) and from the ENSEMBLES project (in the following climate runs 6–13).

Temporal interpolation. Temporal interpolation of measured daily temperature time series was used to obtain hourly time series, following a stepwise procedure of spline interpolation [21]. Resulting hourly temperature time series showed a year-round mean error of -0.031 K h^{-1} and mean absolute error (MAE) of 0.448 C h^{-1} as well as an error of 0.587 hours of frost ($\leq 0°C$) per month of April, compared to measured hourly time series at 56 sites. Time series of the climate model CLM (3 h resolution) were brought to hourly resolution by applying cubic spline interpolation.

Spatial interpolation. Spatial interpolation through ordinary kriging [33] was used to bring measured as well as simulated data to common and regular grids (0.1°·0.1° as well as 0.2°·0.2°) for the area 51° to 54° latitude north and 6.5° to 12° longitude east. While measured data was interpolated directly, simulated hourly temperatures (climate runs 1–5) were previously aggregated by taking the mean of each hour of nine neighboring model grid points (area approximately 30 km·30 km for REMO). By doing so for every model grid point and hence obtaining a spatial floating mean, the original model resolution was maintained. Simulated daily mean and minimum temperature time series were not aggregated due to the coarser spatial resolution.

Bias correction. Since several climate models underestimate the occurrence of frosts, simulated temperature series were bias-corrected for each month by distribution-based quantile mapping [28], using non-parametric transfer functions obtained by applying a Gaussian kernel with bandwidth h = 0.1 [34]. The period of comparison from which transfer functions were derived for bias correction was 54.4±7.3 years for climate runs 1 and 3, 49.8±4.9 years for climate runs 2, 4 and 5 as well as 57.9±4.4 years for climate runs 6–13 (mean ± standard deviation). Hence, the influence of the multidecadal variability was assumed to be negligible. Information on bias correction dynamics with climate runs 6–13 (Table 2) have been published [35].

Projection of temperature. In the following, temperature time series are presented as anomaly from the 1971–2000 mean as indicated by $\Delta T_{y1,y2,s}$, with the centers of the respective periods $y1$ and $y2$ and grid points s (see Methods S1 for equation).

Projection of last spring freeze. The last spring freeze was defined as the last day before July 31st, exhibiting a minimum air temperature $\leq 0°C$, and taken directly for every year from temperature time series.

Phenological Data and Models

Data sources. In order to simulate apple bloom phenophases, time series (Table 2, Figure 1) from the German National

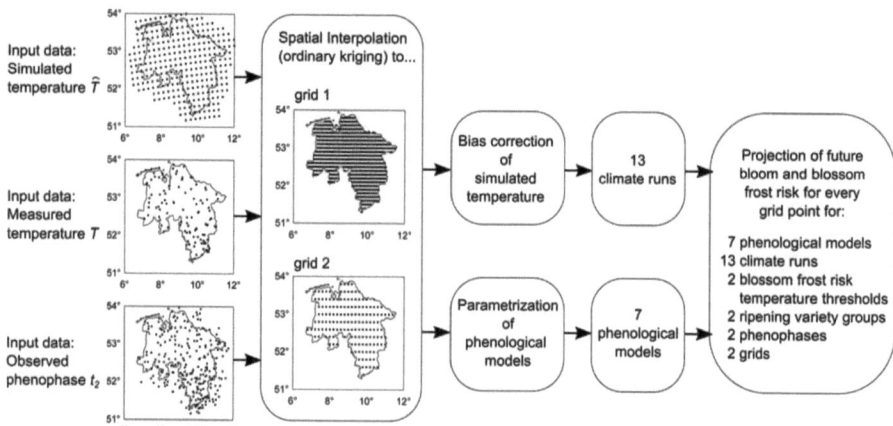

Figure 1. Scheme of used input data and projection. Note that for simulated temperature the grid of the regional climate model CLM is shown exemplarily.
doi:10.1371/journal.pone.0075033.g001

Meteorological Service (htp://www.dwd.de) of observed beginning of flowering (first flowers open) as well as onset of full bloom (50% of flowers open), defined as phenophases 60 and 65 on the BBCH-scale [13], were processed and used to calibrate phenological models for early and late ripening varieties as follows.

Spatial interpolation. Phenological time series were spatially interpolated as described above for measured temperature time series.

Basic phenological models. In principle, all applied phenological models (Table 3, 4, Methods S1) assume that the time of bloom is related to so-called sums of chilling and heat units (Sc, Sf) accumulated during winter (chilling phase) and spring (forcing phase), (see Table 4 for denominations). It is assumed, that Sf is related to Sc [36,37]. The basic models (Table 3, models 1–4) have been described in the literature [17–21] and their equations are given in Methods S1.

Extended phenological models. Models including an additional day-length-parameter for the calculation of the forcing phase were included in the ensemble (Table 3, models 5–7), as a higher performance of model no. 5 has been reported. Models 6–7 are new model variations of the sequential and parallel chilling-forcing models [23], which were extended for a factor for the length of day D, assuming that bloom is influenced by radiation only during the forcing phase. For both, the rate of forcing Rf was calculated as follows:

$$Rf(T_i) = \begin{cases} 0 & \text{if } T_i \leq Tbf \\ \frac{28.4}{1+e^{(-0.185(T_i - Tbf - 18.4))}} \cdot \left(\frac{D}{10}\right)^c & \text{else} \end{cases} \quad \text{with} \quad (1)$$

Rf : Rate of forcing $[-]$

T_i : Daily mean air temperature at day i [°C]

Tbf : Base temperature [°C]

D : Length of the day [h]

c : Calibration parameter $[-]$

Table 3. Phenological models.

No.	Type	Daylength	Tbf	Tbc	Sf(t₁)	Sc(t₁)	t₁	a	b	c	Ref.
1	Thermal time	–	+	–	+	–	+ᵃ	–	–	–	[20]
2	Sequential chilling-forcing	–	+	+	+	+	–	+	+	–	[23]
3	Parallel chilling forcing	–	+	+	+	+	–	+	+	–	[23]
4	Modified Utah	–	+	+	+	+	–	–	–	–	[17,43]
5	Thermal time	+	+	–	+	–	+	–	–	+	[25]
6	Sequential chilling-forcing	+	+	+	+	+	–	+	+	+	–
7	Parallel chilling forcing	+	+	+	+	+	–	+	+	+	–

ᵃFor model 1, t1 was set to January 1.
doi:10.1371/journal.pone.0075033.t003

CHAPTER 3. INVESTIGATIONS   3.3. BLOOM / BLOSSOM FROST RISK

Future Blossom and Frost Risk for *M. domestica*

Table 4. Denomination of variables and parameters.

Notation	Description	Unit
T	Air temperature	°C
Tbc, Tbf	Base temperature for chilling, forcing	°C
t	Time	hour [h], day [d] or year [a]
t_0	Start of the chilling period (dormancy)	day of the year [DOY]
t_1	Chilling requirement completed, start of forcing	day of the year [DOY]
t_2	Forcing completed (BBCH 60, BBCH 65)	day of the year [DOY]
Sc, Sf	State of chilling, state of forcing	–
Rc, Rf	Rate of chilling, rate of forcing	–
D	Daylength	h
a, b, c	Calibration parameters	–
i, s, z	Index variables	–
θ	Blossom frost risk	–
β	Temperature threshold for blossom frost	°C
λ	Parameter for calculation of mean and confidence level	–

doi:10.1371/journal.pone.0075033.t004

Parameter estimation and model validation. Models were parametrized for each grid point by fitting the models to observed bloom (BBCH-scale [13], stages 60 and 65 for early and late ripening varieties of *Malus domestica*) and measured daily air temperature (Table 2). Fitting was performed through bound-constrained simulated annealing, minimizing the root mean square error (RMSE) between observed and simulated day of the year (DOY) of bloom. Simulated annealing for parameter estimation of phenological models has been described in detail [38] and was performed in the present study by using the Global Optimization Toolbox (The Mathworks Inc., Natick, Massachusetts) on a computing cluster system (http://www.rrzn.uni-hannover.de/clustersystem.html). For this, *Tbc* and *Tbf* were searched between 0°C and 10°C, as this is believed to be the effective range of temperature on the development of apple trees [23]. The models were validated internally (same location) as well as externally (different location) by calculating the prediction root mean square error (PRMSE) determined by full-cross validation ("leave-one-out") and by applying the model with optimized parameters to six different and randomly chosen locations in the range of 20 to 28.3 km distance.

All models accounting for *Sc* were initiated with t_0 = 1 August. The simple thermal-time model (1) was started with fixed t_1 = 1 (January 1st, model 1), whereas the extended thermal-time model (5) was started on August 1st (DOY 213, 214) in order to optimize t_1. Models 1 and 5 do not account for a chilling phase and hence implicitly assume that chilling is already completed at t_1.

Projection of Bloom

Bias-corrected air temperature time series of 13 climate realizations (Table 2, Figure 1) were used as input for seven phenological models for 792 locations in Lower Saxony on a 0.1°·0.1° grid (climate runs 1–5) and for 274 locations on a 0.2°·0.2° grid (climate runs 6–13, Table 2, Figure 1) to project future apple bloom. Projections were conducted for all grid points whereas presented results were restricted to the area of Lower Saxony (Figure 1) in order to avoid boundary effects due to interpolation. Comparison of results from all 13 projections took place on the grid of lower resolution. All simulations were conducted with early as well as late ripening varieties and for two phenological stages (BBCH 60, 65). The change in blooming date $\Delta t_2 y1, y2, s$ with the centers of the respective periods $y1$ and $y2$ and grid points s was calculated as the difference in the 30-year-mean for each grid point. Years with unfulfilled chilling were recorded by counting years without bloom or bloom projected for DOY > 200 as fraction of occurrences in a 30-year-mean. Please see Methods S1 for equations.

Projection of Blossom Frost Risk

Subsequently, years with occurrences of frosts (daily minimum temperature ≤0°C) and possibly blossom damaging situations (daily minimum temperature ≤2°C) during the time from simulated bloom (BBCH 60, BBCH 65) to the 31st of July of each year were counted separately. The additional threshold of 2°C was chosen in order to account for spatial discrepancies of observed bloom and measured temperature as well as for possible radiation frosts with tissue temperatures falling below air temperature [19], measured at standard meteorological conditions. Blossom frost risk was defined as the ratio of number of years with temperatures lower or equal to a predefined threshold occurring after a specific phenophase in 30 years:

$$\theta_{y,s} = \frac{1}{30} \cdot \sum_{i=-14}^{15} \mu_{i,s} \text{ with}$$

$$\mu_{i,s} = \begin{cases} 1 & \text{if} \quad min(\{T_{y+i,t_2y,s,s} \ldots T_{y+i,\omega,s}\}) \leq \beta \\ 0 & \text{else} \end{cases}$$

$\theta_{y,s}$: blossom frost risk of year y at grid point s, [−]

$T_{y,d,s}$: array of daily minimum temperature of year y, day d and grid point s [°C]   (2)

β : temperature threshold, either 0 or 2 [°C]

ω : 212 or 213 (leap year) for 31.7., [DOY]

$t_2 y,s$: onset of phenophase, e.g. begin of bloom of year y at grid point s

y : year of calculation, e.g. 1980

i : index

s : grid point

The change in blossom frost risk $\Delta\theta$ was calculated from 30-year-means of each grid point:

$$\Delta\theta_{y1,y2,s} = \theta_{y2,s} - \theta_{y1,s} \text{ with}$$

$\Delta\theta_{y1,y2,s}$: projected change in blossom frost risk from year $y1$ to year $y2$ of every grid point s in Lower Saxony, [−]   (3)

$y1, y2$: year of calculation (past, future)

s : grid point

Probability mass functions were calculated in order to estimate the distribution of changes in blossom frost risk till the end of the 21st century (2070–2099 minus 1971–2000). The values of these probability mass functions were estimated non-parametrically with the help of kernel density estimation, applying a Gaussian kernel. Please see Methods S1 for equations.

Partitioning of Uncertainty of Temperature, Bloom and Blossom Frost Risk

In order to estimate the meaningful projection horizon (= 'Time of emergence', [39]) of the results obtained as described above, the fractional variance of the system was calculated and the total variance of the projection was partitioned. For this purpose the methodology of Hawkins and Sutton [31] was applied to the presented projections for the day of bloom t_2. Instead of looking at different climate models and scenarios, the present work analyzes the internal variability, the uncertainty from climate realizations of one IPCC-scenario (A1B) and the variance resulting from the impact models. Impact models were weighted by their error as described for climate models [31]. The following calculations were carried out with 10 year mean moving average time series of the area mean of Lower Saxony (mean of all grid points s, please see Methods S1 for equations). In brief, the total variance for bloom was calculated as described below. Projection uncertainty of temperature and blossom frost risk was calculated as described for bloom (temperature analysis only for internal and climate realization variability).

$$B_{total}(y) = B_1 + B_2(y) + B_3(y) \text{ with}$$

B_{total} : Total variance of projected bloom, $[d^2]$

B_1 : Internal variability (residual variance), $[d^2]$

B_2 : Uncertainty of climate realizations (variance across climate runs), $[d^2]$ (4)

B_3 : Uncertainty of impact models (variance across phenological models), $[d^2]$

y : year of calculation, e.g. 1980

The contribution of B_1, B_2 and B_3 to the total variance can be expressed as fraction of the total variance:

$$H_z = \frac{B_z \cdot 100}{B_{total}}$$

H : Fraction of the total variance, [%] (5)

z : 1, 2, 3

The mean change in blooming dates of all projections (climate impact signal) over the reference period was obtained as:

$$G(y) = \frac{1}{n} \sum_{s,z} W_s x_{s,z,y} \text{ with}$$

W : model weight, [−]

x : change of phenophase, Δt_2, compared to 1971−2000 [d] (6)

s : impact model (2−7)

z : climate realization (1−13)

n : number or climate realizations, [−]

y : year of calculation, e.g. 1980

Models were weighted (eq. 6) with weights W inversely proportional to their model error (see [31]), giving models with lower errors comparatively more importance. From G and B_{total} the fractional uncertainty F, which is the inverse of the signal-to-noise ratio, was calculated as follows:

$$F(y) = \frac{\lambda \sqrt{B_{total}(y)}}{G(y)} \text{ with}$$

λ : parameter for calculation of confidence levels 50% ($\lambda = 0.67$), 68% ($\lambda = 1$) and 90% ($\lambda = 1.65$) (7)

Statistics of Single Time Series

Continuous time series of calculated completion of dormancy, blooming date and last spring freeze were analyzed using a Mann-Kendall-test [40], whereas trends in blossom frost risk were analyzed with a test by Cox & Lewis [41].

Results

Validation of Methods

The presented methodology was evaluated at the levels climate, quality of phenological model in order to simulate phenophases as well as blossom frost risk. A bias correction had no influence on the mean temperature pattern, whereas the accuracy of simulated frost distribution was drastically improved (Table 5), see also [35]). While climate model time series underestimated frosts in April, this was corrected through the bias correction.

Models could be fitted to reproduce bloom with 3.2 to 5.7 d mean accuracy (RMSE), whereas testing models with fitted parameters (see Methods S1) for different locations revealed an external PRMSE of 3.9 to 8.0 d (Table 6). While the thermal time model (1) exhibited the highest mean error (1.8 d higher than mean of other models), the thermal time model with extension for day length exhibited the lowest mean error (2.0 d lower than mean of other models). On average models (1–3) were improved by 2.0 d when accounting for day length (models 5–7), whereas performance did not differ greatly between BBCH-stages 60 and 65 nor between early and late ripening varieties.

Blossom frost projection accuracy was verified at different levels, since direct comparison of measured blossom frost with blossom frost from simulated time series is not possible in a direct manner

3.3. BLOOM / BLOSSOM FROST RISK

Table 5. Stepwise error of simulation chain segments. SE: Simulation error, ABS: absolute level from measured data.

Parameter		T bias corrected	Frost occurrences per month of April		Bloom[a]	Blossom[b] frost risk θ
			[h]	[d]	[d, DOY]	[−]
Frost	ABS	−	25	4	−	−
Frost	SE[c]	no	7	3	−	−
Frost	SE[c]	yes	<1	<1	−	−
Bloom	ABS	−	−	−	117–126	0.163
Bloom	SE[c]	no	−	−	−	−
Bloom	SE[c]	yes	−	−	4–8	−
Blossom frost	ABS	−	−	−	−	0.163
Blossom frost	SE[c]	no	−	−	−	−
Blossom frost	SE from phenol. models[cd]	yes	−	−	−	0.001–0.034
Blossom frost	SE from time series[ce]	yes	−	−	−	0.021–0.075

[a]min-to-max range across all ripening groups and phenophases.
[b]min-to-max range across all ripening groups, phenophases and phenological models.
[c]Mean absolute error (MAE), average over all grid points.
[d]Error from comparison of measured blossom frost risk with blossom frost risk simulated with measured temperature (1991–2012).
[e]Error from comparison of blossom frost risk simulated with measured temperature with blossom frost risk simulated with simulated temperature (1951–2012).
doi:10.1371/journal.pone.0075033.t005

for short periods (<30 a). Therefore the influences of phenological models and of time series on blossom frost incidents were extracted separately. Applying the phenological models to measured climate data of the calibration period 1991–2012 reproduced blossom frost incidences from measured temperature and measured bloom (Figure 2, Table 5). Subsequently the influence of the time series on blossom frost projection accuracy was tested by applying the validated phenological models on measured and on simulated-bias corrected time series (1951–2012). Despite bias correction, projection with simulated-bias corrected time series showed a mean absolute error (MAE) of blossom frost risk of up to 7.5 percentage points (Table 5). However, mean influences of impact model and time series on blossom frost risk projection accuracy were 1.4 and 3.6 percentage points respectively (mean MAE). Finally blossom frost risk was biased by +0.9 and −3.6 percentage points by impact model and time series, respectively, still resulting in an overall underestimation of blossom frost.

Dormancy and Bloom

In the mean, observed bloom from 1991 to 2012 changed by −3.3 d K^{-1} (R^2=0.87) while air temperature increased by 0.037 K a^{-1}. Phenological models, which were calibrated with these data, gave the following results when applied to simulated temperatures. All chilling-forcing models consistently showed a delay for the release of dormancy t_1 (Figure 3) with major changes not occurring before 2030, following the temperature warming patterns of both simulated climate data sets. However, t_1 showed a larger spread across ENSEMBLES runs than for ECH5-REMO/CLM simulations, while the number of years with unfulfilled chilling requirement increased in both cases (Figure 4). Unlike t_1, projection of the onset of the phenological phases for t_2 (BBCH 60, 65) revealed an advancement. While models 2–7 follow a relatively homogeneous pattern, model 1 projects a faster advance. These main patterns also become visible on a regional scale (Figure 5,6). However, changes in the day of bloom vary regionally depending

Table 6. Prediction Root Mean Squared Error PRMSE of phenological models [d].

Model	early ripeners		late ripeners		mean
	BBCH 60	BBCH 65	BBCH 60	BBCH 65	
1	7.97	7.26	7.28	7.27	7.45
2	6.67	5.95	6.24	6.03	6.22
3	7.10	6.30	6.54	6.25	6.55
4	6.81	6.83	6.54	6.67	6.71
5	4.14	4.12	3.91	4.34	4.13
6	4.96	5.08	4.88	5.10	5.00
7	5.13	5.19	4.89	5.29	5.13
mean	6.11	5.82	5.75	5.85	5.88

doi:10.1371/journal.pone.0075033.t006

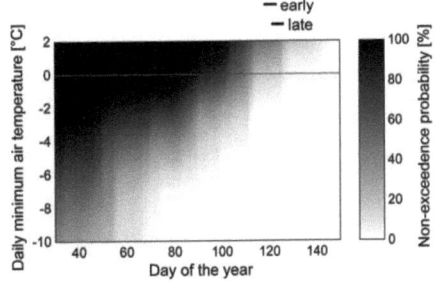

Figure 2. Present temperature incidence of Lower Saxony (1991–2010). Bars indicate mean flowering period (BBCH 60–65) of early and late ripening varieties.
doi:10.1371/journal.pone.0075033.g002

3.3. BLOOM / BLOSSOM FROST RISK

Future Blossom and Frost Risk for *M. domestica*

Change in yearly mean temperature

Change in fulfillment of chilling requirement

Change in blooming date

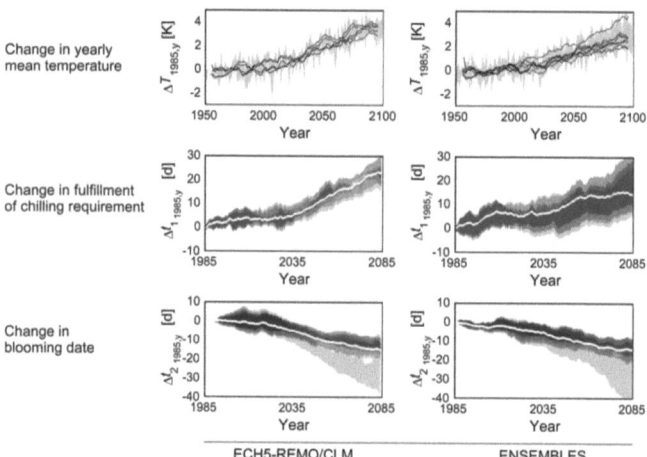

ECH5-REMO/CLM   ENSEMBLES

Figure 3. Projected changes in air temperature, fulfillment of chilling requirement and onset of flowering. Projected with 5 (ECH5-REMO/CLM) and 8 (ENSEMBLES) climate runs and five (Δt_1) and seven (Δt_2) phenological models for Lower Saxony (area mean), relative to the 1971–2000 mean. ΔT: single year-mean, min-to-max range of climate runs (shaded area), 10 year moving average of each run (solid lines, see Methods S1 for equation). Δt_1, Δt_2: BBCH 65, early ripeners, 30-year-moving-average, all impact model mean (solid white line), single model range (shaded areas). The range of each phenological model (min-to-max) obtained from climate runs is plotted with 20% transparency (darker areas illustrate coinciding results).
doi:10.1371/journal.pone.0075033.g003

on the model. Regarding the timescale, all models project a shift in the day of bloom of −5.4±3.0 d by 2035 compared to 1971–2000 (area mean, all varieties and stages), whereas results for 2084 differ. While model 1 shows the strongest change (−26.7±8.2 d), models 2–7 project a mean shift of approx. −12.9±3.3 days. The latter again differ in their regional variation. Although the classic sequential and parallel chilling forcing models (2–3) show a similar mean shift of bloom as their versions extended for daylength (models 5–7; −13.5 d und −11.2 d respectively), the former exhibit higher variation (±3.6 d and ±2.2 d respectively). A similar variation was also found for model 4 (±3.3 d).

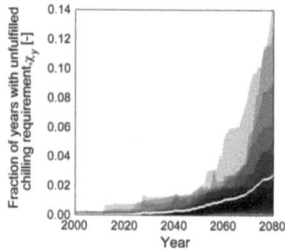

Figure 4. Proportion of years with unfulfilled chilling requirement. Areas: min-to-max range across seven phenological models for each climate run (area mean of Lower Saxony, 30-year moving average); white line: Mean of impact models and climate runs.
doi:10.1371/journal.pone.0075033.g004

Projected Last Spring Freeze and Blossom Frost Risk

According to the scenario and climate runs considered, the last spring freeze (≤0°C) will shift by −10.0±4.2 days and −27.3±7.4 days by 2035 and 2084 respectively, with regard to the reference period 1971–2000 (Figure 7). Hence these 30-year-mean trends indicate an increasing discrepancy of the day of bloom and the last spring freeze. Correspondingly the mean occurrences of blossom frost (θ) are projected to decrease in the long run (Figure 5,6). Nevertheless model 1, which showed the fastest advancement of bloom, projects a mean increase of blossom frost risk by 3.4 percentage points whereas models 2–7 project a mean change by −4.1± 3.3 percentage points, ranging from −2.6 percentage points for late ripeners (BBCH 65) to −6.0 percentage points for early ripeners (BBCH 60). In the mean, runs of EH5-REMO/CLM and ENSEMBLES runs produced similar estimates for changes in blossom frost risk (−2.7±4.4 percentage points and −3.2±4.5 percentage points respectively). However, all models also exhibited regional and temporary increases in blossom frost occurrences. The resulting probability mass function values (pmf) are shown in Figure 8, displaying also the contrary result of model 1. A larger spread and stronger decrease was observed for the probability of temperatures of ≤2°C after onset of phenophases.

Projection Uncertainty

Phenophases followed temperature patterns closely, with early and late ripening varieties advancing at 5.6 and 5.4 d K^{-1} respectively and BBCH 60 and BBCH 65 advancing at 5.6 and 5.4 d K^{-1} respectively, resulting in a mean change of −5.5 d K^{-1} (Figure 9). Higher correlations were found between changes in begin of flowering date and mean temperatures between February and April (−6.1 d K^{-1}, R^2 = 0.93). However no correlation was

CHAPTER 3. INVESTIGATIONS   3.3. BLOOM / BLOSSOM FROST RISK

Future Blossom and Frost Risk for *M. domestica*

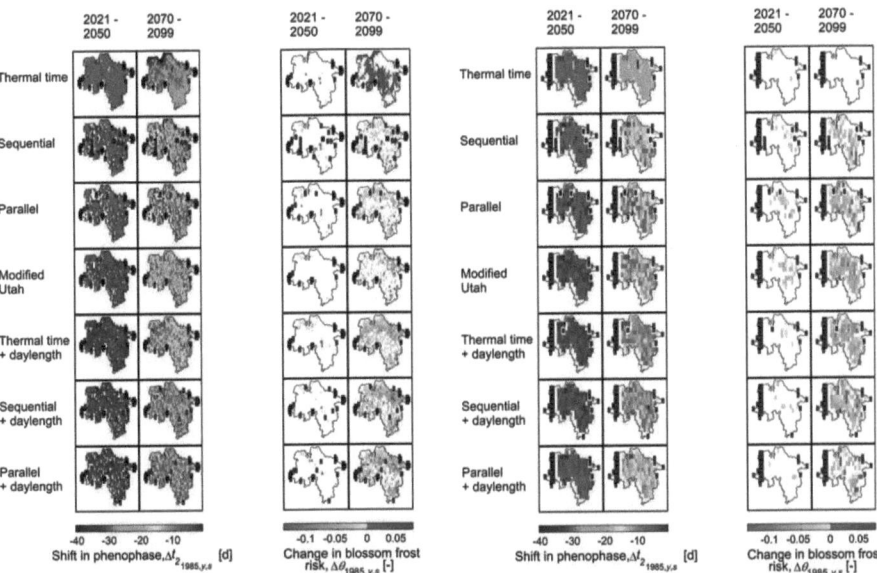

Figure 5. Changes in bloom and blossom frost risk as projected by different phenological models and climate runs 1–5. Early ripeners, BBCH 65, temperature threshold $\beta = 0°C$, reference period 1971–2000, resolution 0.1°. White fields denote non-significant results, black fields denote missing/insufficient data. 1–99% percentile range. $y = 1985$ and 2084, $s =$ grid point.
doi:10.1371/journal.pone.0075033.g005

Figure 6. Changes in bloom and blossom frost risk as projected by different phenological models and climate runs 6–13. Early ripeners, BBCH 65, temperature threshold $\beta = 0°C$, reference period 1971–2000, resolution 0.2°. White fields denote non-significant results, black fields denote missing/insufficient data. 1–99% percentile range. $y = 1985$ and 2084, $s =$ grid point.
doi:10.1371/journal.pone.0075033.g006

found between changes in the respective variances of temperature and flowering dates, with exception of the simple thermal time model (model 1, data not shown).

The projection uncertainty increased with increasing lead time (Figure 10, top) and for the period investigated, the accuracy of the projection of t_2 in the short run is mainly dependent on the projected climate and internal variability. With increasing horizon of projection, the climate signal (temperature) becomes stable while impact/phenological model results diverge. Consistently fractions of climate and internal variability of the total variance decreased with increasing lead time (Figure 10, bottom). Finally, the projection accuracy at the end of projection horizon depended equally on the climate and impact/phenological model variance.

The resulting fractional uncertainty F decreased over time. Comparing the sources of uncertainty, the fractional uncertainty of temperature time series decreased faster than of blooming date and blossom frost risk time series. Accordingly, the lowest level of fractional uncertainty at any of the confidence levels investigated was also reached by temperature. While the 90% percentile for temperature and bloom reached 1 in 2019 and 2042–2044 respectively, the uncertainty of blossom frost risk passed 1 only by the 68% percentile (±1 standard deviation) by 2077 (Figure 11). From this point on, the projected change (signal) exceeded the variance of the projection (noise). A minimum of the fractional uncertainty was found for 2078 (temperature), 2083–2084 (bloom) and 2085–2088 (blossom frost risk), after which it was projected to

increase. This result was similar for early as well as late ripening varieties and for both BBCH stages.

Discussion

Phenological Models

Projections with pure forcing models [20,21] are subject to changes in dormancy completion [23] and varying warming of the seasons. The application of such a model in the present study produced similar results of increasing risk as in the mentioned literature, but different to the main outcome of the present

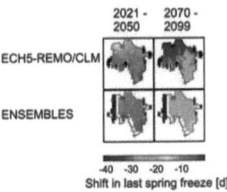

Figure 7. Changes in last spring freeze. Reference period: 1971–2000. White fields denote non-significant results, black fields denote missing/insufficient data. 1–99% percentile range.
doi:10.1371/journal.pone.0075033.g007

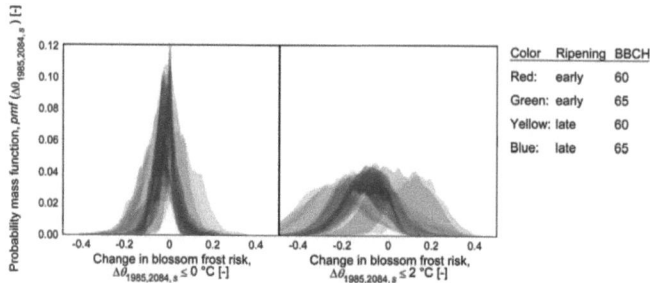

Figure 8. Distribution of projected changes in blossom frost risk by the end of the 21st century (2070–2099 minus 1971–2000) for early and late ripening varieties, phenophases BBCH 60 and 65 and 7 phenological models: Temperature thresholds ≤0°C and ≤2°C; inter-quartile range across 13 climate runs; phenological models are presented by same colors. Calculated from all grid points s (see Methods S1 for equation).
doi:10.1371/journal.pone.0075033.g008

ensemble study. For this reason, sequential or parallel chilling-forcing models have been recommended [23], as well as models including nearly time-invariant factors as day length [25]. The mean error of all models presented (5.9 d) was in the range of published model performances [15,20,21,23,25,42,43]. This error must be seen in context to the observed flowering duration (BBCH 60 to BBCH 67), which ranged during the calibration period from 6 to 27 d (1 to 99% percentile range). As large errors in simulated flowering dates can erroneously increase the blossom frost risk, the influence of the RMSE on the simulated blossom frost risk was tested (not shown), but no significant influence was found in the range of the calibrated models errors. Having further a negligible bias, the models were rated as suitable for blossom frost risk projections from this point of view. Furthermore, in the present work models were improved by including day length, thus confirming previous findings [25]. Also other models including

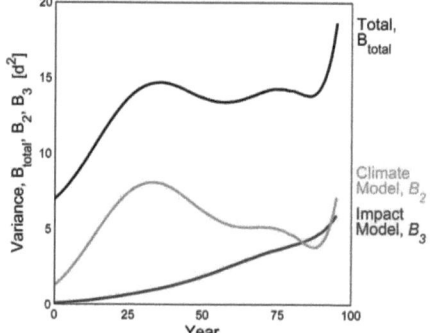

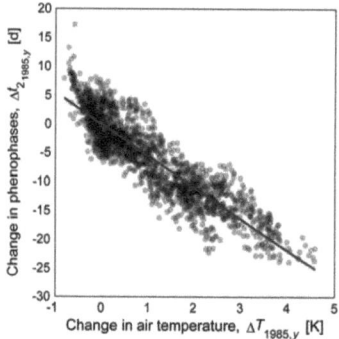

Figure 9. Simulated relation between projected absolute changes in decadal mean air temperature and changes in the day of bloom compared to the 1971–2000 mean. Depicted values are related to 139 years (y = {1956 .. 2094}, see Methods S1 for equation) and 13 climate realizations for the area mean of 2 phenophases and 2 variety groups. Slope of regression (solid line) = −5.4842, offset = 0.0385, R^2 = 0.81.
doi:10.1371/journal.pone.0075033.g009

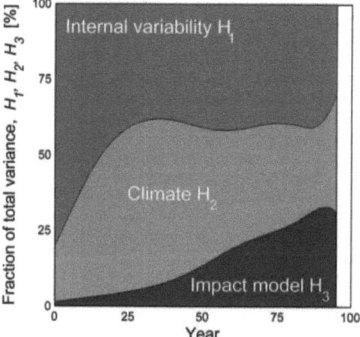

Figure 10. Uncertainty in the projection of apple bloom (t_2). Drawn from phenological impact models 2–7 and 13 climate projections. Mean uncertainty of phenophases (BBCH 60, 65) and ripening groups (early, late).
doi:10.1371/journal.pone.0075033.g010

3.3. BLOOM / BLOSSOM FROST RISK

Future Blossom and Frost Risk for *M. domestica*

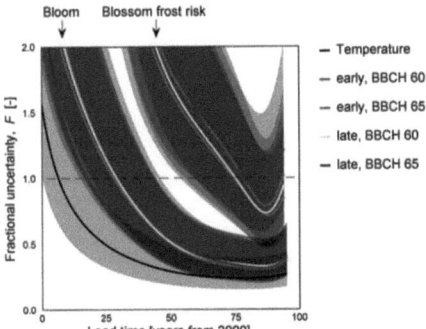

Figure 11. Uncertainty pattern of projected temperature (T), apple bloom (t_2) and apple blossom frost risk (θ). 68.3% percentile (solid lines) and 50-to-90% percentile ranges (gray areas) from 13 climate projections and phenological impact models 2-7 (bloom, blossom frost risk).
doi:10.1371/journal.pone.0075033.g011

exponential terms were applied in blossom frost risk estimation [17,43], relying solely on temperature as input. As they increase the "resistance" for each computation of a day of the year for flowering, exponential models eliminate one deficit of pure temperature sum models which is a calculated flowering date beyond summer in exceptional years, leading to high errors (given that dormancy is completed). In addition, the error of models including a parameter for day length might be lower due to a higher number of parameters. This statistical effect should be separated from the physiological meaning of the parameter. As the role of the length of day in flowering physiology of apple is still under debate [26], these model properties cannot be isolated for the present study, but should be regarded in the future. Finally, while presented combination of sequential or parallel models with an exponential term for day length improves model robustness, these models are also more complex.

Influence of Climate Change on the Onset of Phenophases

The observed effects of delayed completion of the chilling requirement and earlier flowering due to faster completion of heat requirement are well known[6,15,42,44–46]. Thereby the extension of the growing season [47,48] and the advancement of flowering dates during the past due to climate change have been studied largely for several tree species[44,49–51] including apple flowering phenology [9,14,42], allowing the assumption of a general trend. Accordingly "very similar" reactions of apple and cherry blossoming (BBCH 60) as well as winter rye stem elongation (BBCH 31) to early spring conditions were observed [14]. However, the observed mean change of onset of flowering (BBCH 60) of -3.3 d K^{-1} during the short calibration period of phenological models (1991–2012) were lower than those reported from other studies for the entire second half of the 20th century. These published estimates range from -7 to -8 d K^{-1} of year-mean temperatures (values calculated from [9,42]) for late ripeners up to -5 d K^{-1} of mean temperatures from February to April [14] for early ripeners. Still these discrepancies should result from geographic and orographic differences from the present to the

mentioned publications: Analyzing the present model projections for the same periods as in the mentioned literature (1958–2007, 1976–2002, 1969–1998) fairly reproduced these dependencies with -7.5 up to -8.6 d K^{-1} for late, and -6.5 d K^{-1} (February-April temperatures). Consistently, also the projected findings for changes in the onset of apple flowering of -5.4 to -5.6 d K^{-1} (all varieties and stages and years) and -6 d K^{-1} (BBCH 60, February-April temperatures) are in a comparable range. From this can be concluded, that apple flowering phenophases have a clear and comparable reaction to changes in temperature despite differences in region and varieties and that this impact can be tracked by one-dimensional phenological models in combination with climate ensembles.

Furthermore, despite a continuous advancement of flowering dates, an opposing effect of delayed release of dormancy and enhanced spring warming was observed. While warmer winters result in reduced chilling, they can be compensated to a certain extent by warmer springs [52]. For apple bloom this has been reported for the past [42]. However, reduced chilling will eventually slow down the advancement of flowering dates as postulated [42,52] and as deduced from the relative changes for t_1 and t_2 in the present study for the 2nd half of the 21st century. In addition, eventually years with unfulfilled dormancy will occur. Such events have not been observed in Germany during the past century [6], but are discussed for the future [6,45,46]. A rough estimate for the probability of years with unfulfilled chilling requirement of up to 15% can be found for the largest producing area in Lower Saxony (Niederelbe) [53]. While this estimate coincides with the here presented range, the mean fraction of years with unfulfilled chilling requirement is lower (3.7%). Following the authors, it must be stated, that these projections are subject to large uncertainties and require further investigation.

Spring Freeze and Blossom Frost Risk

Last spring freeze follows the warming pattern with changes of increasing speed towards the second half of the 21st century. The projected shifts for the period 1985–2035 (30-year-means) of -2.0 d/decade are in the range of those changes reported for the second half of the 20th century for Central Europe (-2.2 d/decade [12]). Following the future warming pattern in simulations, last spring freeze is likely to change about -3.5 d/decade (2035–2084).

Blossom frost risk possibly decreases in the long term. This result can be obtained roughly by putting together the relative advancement of projected bloom and last spring freezes, as well as in more detail through the present computation with single models. Starting with a blossom frost risk of up to 16%, simulations showed a decline in blossom frost occurrence to about half by the end of the 21st century. Nevertheless, blossom frost is unlikely to disappear and staying at a comparable level as present until the middle of the century. As blossom frost risk strongly depends on the region, period, variety and BBCH stages, publications are hardly comparable. While the present observations and computations for the past are in the range of other studies [9,19,20], projected results differ. The often stated hypothesis of an increase in blossom frost risk due to advanced bloom in combination with increased variance in the last spring freeze date [19] does not hold true for the present study, as spring freezes declined comparably faster than flowering dates.

Projection Uncertainty

Climate impact projection to a near future is often highly uncertain since the internal variability of the system at hand is larger than the expected changes at point of time. As these changes increase with time and relatively to the total variance of the projection, more confidence in the projection signal is gained. Future climate is commonly assessed in ensemble run projections, including RCMs [54] and bias-corrected simulations [35]. Sampling, climate model, radiative and boundary uncertainties have been investigated for climate models, varying for RCMs across field, region and season [54]. While such climate ensembles are also increasingly used to drive impact models [55], the impact models error adds to the signal strength. Uncertainty of climate projections increases with increasing simulation members, as clearly shown by the different patterns of fractional uncertainty of temperature and bloom as well as blossom frost risk. Thereby projection uncertainty of surface temperature depended only on the different climate models, whereas bloom depended on climate and impact models and blossom frost risk additionally depended on the interaction of projected bloom and temperature.

In the present approach times of emergence of 34 years and 57 to 59 years were estimated for temperature and blooming date respectively (compared to the mean 1971–2000), considering one SRES scenario (A1B). This is in the range of the estimated time of emergence for regional surface temperatures of SRES scenarios A2, A1B and B1 from GCMs [39]. While the approach relies heavily on the chosen climate ensemble and impact models, larger variance can be expected with increasing spatial (or temporal) resolution. Therefore the estimated lead time for the minimum of uncertainty of ∼100 years (2078–2088) is consistent with ∼30 to 80 years established for temperature [31]. However, the present works investigated a range of climate and impact models of one scenario, while the cited publications investigated three scenarios for climate models. Hence further projections of future bloom are required in order to remove this lack of comparability. Nonetheless, looking at the cooler scenario B1 and neglecting the similar scenario A2 for central Europe, a larger spread in the day of bloom and hence in the estimated blossom frost risk can be expected, increasing the time of emergence of the climate impact signal. Transferring the estimated time of emergence to other climate impact studies from different research fields by assuming similar variability across models would imply, that a large fraction of these studies operates at the very edge of statistical significance. For example, from a review on 14 publications on future risks through wheat diseases [56], 8 include statements and 2 are solely based on statements for a time horizon ≤2030. From the present findings, the statistical meaning of these studies must be carefully put into context.

Two effects arise: On the one hand, using a location parameter (e.g. mean or median) of a climate ensemble as input for impact models may produce significant future changes while ignoring climate projection uncertainty. On the other hand, using single impact models and/or fixed impact model parameters can give only mean tendencies, similarly ignoring parameter ranges in climate impact. The presented results show these effects, as single impact models with climate ensemble mean as input show consistently significant trends of advancing bloom and, with one exception, of decreasing blossom frost risk. Regarding the joint uncertainty of climate and impact models, this may hold true for bloom beyond the estimated projection horizon. However, projected changes in blossom frost risk are low compared to the variability across models. While this is a particularly pronounced problem of extreme events such as blossom frost, it has severe consequences. From the present results, despite a tendency of decreasing blossom frost risk, it must only be concluded that future blossom frost risk is very unlikely to increase.

Limitations

The present work does not consider the severity and distribution of frosts. Hence it must be taken into account, that other plant reactions than those investigated and resulting from frost distributions may dominate in the future. As actual blossom frost damages were not evaluated, the presented results depict the blossom frost risk tendency. Although blossom frost damage severity increases with decreasing temperature [5], temperatures cannot be translated directly into economic losses, as frost protection (e.g. sprinkler) takes place in practice. Furthermore employed models accounted for day length, but did not use actual surface radiation from climate models. Hence possible effects due to changes in light conditions (e.g. phenological effects) and effects due to severe radiation (radiation frosts) are not represented to full extent. Additionally, the influence of the day length on apple flowering physiology remains uncertain. Despite low availability of consistently bias corrected climate time series of high temporal resolution [29], future approaches should consider this. Finally, future changes in varieties were not taken into account albeit varieties might respond differently to blossom frost [57].

Conclusions

Regarding the aspects of phenological model structure, simulation uncertainty as well as blossom frost risk, the following conclusions must be drawn from the present findings. Despite a lack of physiological explanation, phenological model performance is improved by including the length of the day. However, projection results from single time series must be put into context to the uncertainty of the modeling chain, considering the significant projection horizon. The latter depends on the investigated variable and was determined for the present simulation of bloom at 2042–2044. Differently, a minimum of uncertainty was estimated for temperature, bloom and blossom frost risk for the range 2078–2088. Finally the resulting regional blossom frost risk cannot be expected to increase in the long term, as compensatory effects of delayed fulfillment of chilling requirement and faster completion of the forcing phase in spring take place.

Supporting Information

Methods S1 Basic equations and model formulation. (PDF)

Acknowledgments

We kindly thank the RRZN cluster system team at the Leibniz Universität Hannover, Germany, for their support in the production of this work.

Author Contributions

Conceived and designed the experiments: HH. Performed the experiments: HH. Analyzed the data: HH. Contributed reagents/materials/analysis tools: HH. Wrote the paper: HH TR. Review: TR.

References

1. Rodrigo J (2000) Spring frosts in deciduous fruit trees – morphological damage and flower hardiness. Sci Hortic-Amsterdam 85: 155–173.
2. Farajzadeh M, Rahimi M, Kamali G, Mavrommatis T (2009) Modelling apple tree bud burst and frost risk in iran. Meteorol Appl 17: 45–52.
3. Sunley R, Atkinson C, Jones H (2006) Chill unit models and recent changes in the occurrence of winter chill and spring frost in the united kingdom. J Hortic Sci Biotech 81: 949–958.

4. Sugiura T (2010) Characteristics of responses of fruit trees to climate changes in japan. Acta Hortic 872: 85–88.
5. Asakura T, Sugiura H, Sakamoto D, Sugiura T, Gemma H (2011) Frost risk evaluation in apple by modelling phenological changes in critical temperatures. Acta Hortic 919: 65–70.
6. Luedeling E, Blanke M, Gebauer J (2009) Auswirkungen des Klimawandels auf die Verfügbarkeit von Kältewirkungen (Chilling) für Obstgehölze in Deutschland. Erwerbs-Obstbau 51: 81–94.
7. Chitu E, Topor E, Palitineanu C, Dumitru M, Sumedrea D, et al. (2011) Phenological and climatic modelling of the late frost damage in apricot orchards under the changing climatic conditions of south-eastern romania. Acta Hortic 919: 57–64.
8. Fan X, Wang W, Yang X, Wu Y (2010) Responses of apple tree's phonology in east and west sides of liupanshan mountain to climate change. Chinese J Ecol 29: 50–54.
9. Blanke M, Kunz A (2009) Einfluss rezenter Klimaveränderungen auf die Phänologie bei Kernobst am Standort Klein-Altendorf - anhand 50-jähriger Aufzeichnungen. Erwerbs-Obstbau 51: 101–114.
10. Blanke M, Kunz A (2011) Effects of climate change on pome fruit phenology and precipitation. Acta Hortic 922: 381–386.
11. Kunz A, Blanke M (2011) Effects of global climate change on apple 'golden delicious' phenology – based on 50 years of meteorological and phenological data in klein-altendorf. Acta Hortic 903: 1121–1126.
12. Scheifinger H, Menzel A, Koch E, Peter C (2003) Trends in spring time frost events and phenological dates in central europe. Theor Appl Climatol 74: 41–51.
13. Meier U, Graf H, Hack H, Hess M, Kennel W, et al. (1994) Phänologische Entwicklungsstadien des Kernobstes (*Malus domestica Borkh.* und *Pyrus communis* L.), des Steinobstes (*Prunus*-Arten), der Johannisbeere (*Ribes*-Arten) und der Erdbeere (*Fragaria × ananassa Duch.*). Nachrichtenbl Deut Pflanzenschutzd 46: 141–153.
14. Chmielewski F, Müller A, Bruns E (2004) Climate changes and trends in phenology of fruit trees and field crops in Germany, 1961–2000. Agric For Meteorol 121: 69–78.
15. Luedeling E (2012) Climate change impacts on winter chill for temperate fruit and nut production: A review. Sc Hortic-Amsterdam 144: 218–229.
16. Solomon S, Quin D, Manning M, Chen Z, Marquis M, et al. (2007) Contribution of Working Group I to the Fourth Assessment Report of the Intergovernmental Panel on Climate Change. Cambridge University Press, Cambridge, United Kingdom and New York, NY, USA. Available: http://www.ipcc.ch/. Accessed 20 April 2013.
17. Eccel E, Rea R, Caffarra A, Crisci A (2009) Risk of spring frost to apple production under future climate scenarios: the role of phenological acclimation. Int J Biometeorol 53: 273–286.
18. Kaukoranta T, Tahvonen R, Ylämäki A (2010) Climatic potential and risks for apple growing by 2040. Agr Food Sci 19: 144–159.
19. Cannell M, Smith R (1986) Climatic warming, spring budburst and frost damage on trees. J Appl Ecol 23: 177–191.
20. Chmielewski F, Müller A, Küchler W (2005) Climate changes and frost hazard for fruit trees. Annalen der Meteorologie 41 2: 488–491.
21. Hoffmann H, Langner F, Rath T (2012) Simulating the influence of climate warming on future spring frost risk in northern german fruit production. Acta Hortic 957: 289–296.
22. Chmielewski F, Müller A (2005) Possible impacts of climate change on natural vegetation in Saxony (Germany). Int J Biometeorol 50: 96–104.
23. Chmielewski F, Blümel K, Henniges Y, Blanke M, Weber R, et al. (2011) Phenological models for the beginning of apple blossom in Germany. Met Z 20: 487–496.
24. Hökkinen R, Linkosalo T, Hari P (1998) Effects of dormancy and environmental factors on timing of bud burst in *Betula pendula*. Tree Physiol 18: 707–712.
25. Blümel K, Chmielewski F (2012) Shortcomings of classical phenological forcing models and a way to overcome them. Agric For Meteorol 164: 10–19.
26. Körner C, Basler D (2010) Warming, photoperiods and tree phenology. Science 329: 277–278.
27. Nakicenovic N, Alcamo J, Davis G, de Vries B, Fenhann J, et al. (2000) Special report on emission scenarios. Cambridge University Press, Cambridge, United Kingdom and New York, NY, USA. Available: http://www.ipcc.ch/. Accessed 20 April 2013.
28. Piani C, Haerter J, Coppola E (2010) Statistical bias correction for daily precipitation in regional climate models over europe. Theor Appl Climatol 99: 187–192.
29. Hoffmann H, Rath T (2012) Meteorologically consistent bias correction of simulated climate time series for agricultural models. Theor Appl Climatol 110: 129–141.
30. Hoffmann H, Rath T (2012) High resolved simulation of climate change impact on greenhouse energy consumption in Germany. Eur J Hortic Sci 77: 241–248.
31. Hawkins E, Sutton R (2009) The potential to narrow uncertainty in regional climate predictions. B Am Meteorol Soc 90: 1095–1107.
32. Yip S, Ferro CAT, Stephenson DB, Hawkins E (2011) A simple, coherent framework for partitioning uncertainty in climate predictions. J Climate 24: 4634–4643.
33. Oliver M, Webster R (1990) Kriging: a method of interpolation for geographical information systems. Int J Geogr Inf Syst 4: 313–332.
34. Bowman A, Azzalini A (1997) Applied Smoothing Techniques for Data Analysis. New York: Oxford University Press.
35. Dosio A, Paruolo P, Rojas R (2012) Bias correction of the ensembles high resolution climate change projections for use by impact models: Analysis of the climate change signal. J Geophys Res Atmos 117.
36. Landsberg J (1974) Apple fruit bud development and growth; analysis and an empirical model. Ann Bot-London 28: 1013–1023.
37. Murray M, Cannell G, Smith R (1989) Date of budburst of fifteen tree species in britain following climatic warming. J Appl Ecol 26: 693–700.
38. Chuine I, Cour P, Rousseau D (1998) Fitting models predicting dates of flowering of temperate-zone trees using simulated annealing. Plant Cell Environ 21: 455–466.
39. Hawkins E, Sutton R (2012) Time of emergence of climate signals. Geophys res lett 39: 1–7.
40. Mann H (1945) Nonparametric tests against trend. Econometrica 13: 245–259.
41. Cox D, Lewis P (1966) The Statistical Analysis of Series of Events. London: Methuen & Co. Ltd.
42. Legave J, Farrera I, Almeras T, Calleja M (2008) Selecting models of apple flowering time and understanding how global warming has had an impact on this trait. J Hortic Sci Biotech 83: 76–84.
43. Rea R, Eccel E (2006) Phenological models for blooming of apple in a mountaineous region. Int J Biometeorol 51: 1–16.
44. Schwartz M, Ahas R, Aasa A (2006) Onset of spring starting earlier across the northern hemisphere. Glob Change Biol 12: 343–351.
45. Luedeling E, Zhang M, Luedeling V, Girvetz E (2009) Sensitivity of winter chill models for fruit and nut trees to climatic changes expected in california's central valley. Agric Ecosys Environ 133: 23–31.
46. Luedeling E, Zhang M, Girvetz EH (2009) Climatic changes lead to declining winter chill for fruit and nut trees in california during 1950–2099. PLoS ONE 4: 1–9.
47. Chmielewski F, Rötzer T (2001) Response of tree phenology to climate change across europe. Agric For Meteorol 108: 101–112.
48. Tooke F, Battey N (2010) Temperate flowering phenology. J Exp Bot 61: 2853–2862.
49. Menzel A, Sparks T, Estrella N, Koch E, Aasas O, et al. (2006) European phenological response to climate change matches the warming pattern. Glob Change Biol 12: 1969–1976.
50. Ibáñez I, Primack R, Miller-Rushing A, Ellwood E, Higuchi H, et al. (2011) Forecasting phenology under global warming. Phil Trans R Soc 365: 3247–3260.
51. Jie B, Quansheng G, Junhu D (2011) The response of first flowering dates to abrupt climate change in beijing. Adv atmos sci 28: 564–572.
52. Harrington C, Gould P, StClair J (2010) Modeling the effects of winter environment on dormancy release of douglas-fir. Forest Ecol Manag 259: 798–808.
53. Chmielewski F, Görgens M, Kemfert C (2009) KliO: Klimawandel und Obstbau in Deutschland. Abschlussbericht. Humboldt University of Berlin, Institute of Crop Sciences, Subdivision of Agricultural Meteorology. Available: http://www.agrar.hu-berlin.de/fakultaet/departments/dntw/agrarmet/forschung/fp/AB-HU.pdf. Accessed: 20 April 2013.
54. Dqu M, Rowell DP, Lthi D, Giorgi F, Christensen JH, et al. (2007) An intercomparison of regional climate models for europe: Assessing uncertainties in model projections. Climatic Change 81: 53–70.
55. Rojas R, Feyen L, Bianchi A, Dosio A (2012) Assessment of future flood hazard in europe using a large ensemble of bias-corrected regional climate simulations. J Geophys Res Atmos 117.
56. Juroszek P, von Tiedemann A (2012) Climate change and potential future risks through wheat diseases: a review. Eur J Plant Pathol : 1–13.
57. Rugienius R, Siksnianas T, Gelvonauskiene D, Staniene G, Sasnauskas A, et al. (2009) Evaluation of genetic resources of fruit crops as donors of cold and disease resistance in lithuania. Acta Hortic 825: 117–124.
58. Jacob D (2005) REMO A1B scenario run, UBA project, 0.088 degree resolution, run no. 006211, 1h data. cera-db '"REMO– UBA –A1B –1-R006211–1H"'. World Data Center for Climate. Available: http://cera-www.dkrz.de/WDCC/ui/Compact.jsp?acronym=REMO_UBA_A1B_1_R006211_1H. Accessed: 20 Apr 2013.
59. Jacob D, Nilson E, Tomassini L, Bülow K (2009) REMO A1B scenario run, BFG project, 0.088 degree resolution, 1h values. cera-db "remo– bfg– a1b–1h". World Data Center for Climate. Available: http://cera-www.dkrz.de/WDCC/ui/Compact.jsp?acronym=REMO_BFG_A1B_1H. Accessed: 20 Apr 2013.
60. Keuler K, Lautenschlager M, Wunram C, Keup-Thiel E, Schubert M, et al. (2009) Climate simulation with CLM, scenario A1B run no.1, data stream 2: European region MPI-M/MaD. World Data Center for Climate. DOI:10.1594/WDCC/CLM_A1B_1_D2. Available: http://dx.doi.org/10.1594/WDCC/CLM_A1B_1_D2. Accessed: 20 Apr 2013.
61. Keuler K, Lautenschlager M, Wunram C, Keup-Thiel E, Schubert M, et al. (2009) Climate simulation with CLM, scenario A1B run no.2, data stream 2: European region MPI-M/MaD. World Data Center for Climate. DOI:10.1594/WDCC/CLM_A1B_2_D2. Available: http://dx.doi.org/10.1594/WDCC/CLM_A1B_2_D2. Accessed: 20 Apr 2013.
62. van der Linden P, Mitchell J (2009) Ensembles: Climate change and its impacts: Summary of research and results from the ensembles project, technical report. Met Off Hadley Cent., Exeter, U.K. Available: http://ensembles-eu.metoffice.com/. Accessed 20 April 2013.

Future bloom and blossom frost risk for *Malus domestica* considering climate model and impact model uncertainties

Holger Hoffmann[1]*, Thomas Rath,
Biosystems Engineering, Institute for Biological Production Systems, Leibniz Universität Hannover, Hannover, Germany
[1] Present address: Institute of Crop Science and Resource Conservation, University of Bonn, Bonn, Germany
* E-mail: Corresponding hhoffmann@uni-bonn.de

Phenological models supplementary S1

Projection of temperature

Temperature time series are presented as anomaly from the 1971-2000 mean as indicated by ΔT. The anomaly of single years as well as of 10 year moving average time series is shown. By way of example, the latter was calculated as:

$$\Delta T_{y1,y2,s} = \frac{1}{10}\sum_{i=-4}^{5}\frac{1}{n}\sum_{d=1}^{n}T_{y2+i,d,s} - \frac{1}{30}\sum_{i=-14}^{15}\frac{1}{n}\sum_{d=1}^{n}T_{y1+i,d,s} \text{ with} \tag{1}$$

$\Delta T_{y1,y2,s}$: projected change in year-mean air temperature from year $y1$ to year $y2$ of every grid point s in Lower Saxony, [-]
$y1, y2$: year of calculation (past, future)
s : grid point
i : index
d : day
n : number of days of the year (365 or 366)

Projection of bloom

The change in blooming date Δt_2 was calculated as the difference in the 30-year-mean for each grid point:

$$\Delta t_{2\,y1,y2,s} = \frac{1}{30}\cdot\sum_{i=-14}^{15} t_{2\,y2+i,s} - \frac{1}{30}\cdot\sum_{i=-14}^{15} t_{2\,y1+i,s} \text{ with} \tag{2}$$

$\Delta t_{2\,y1,y2,s}$: projected change in blooming date t_2 from year $y1$ to year $y2$ of every grid point s in Lower Saxony, [-]
$y1, y2$: year of calculation (past, future)
s : grid point
i : index

Years with unfulfilled chilling were recorded by counting years without bloom or bloom projected for DOY> 200 as fraction of occurrences in a 30-year-mean:

$$\chi_y = \frac{1}{30} \cdot \sum_{i=-14}^{15} \mu_i \text{ with} \tag{3}$$

$$\mu_i = \begin{cases} 1 & \text{if} \quad t_{2,y+i} > 200 \\ 0 & \text{else} \end{cases}$$

χ : Fraction of years with unfulfilled chilling requirement, [-]
$t_{2,y}$: onset of phenophase in year y, [DOY]
y : year of calculation, e.g. 1980
i : index

Calculation of probability mass functions

The values of probability mass functions were estimated non-parametrically by applying a Gaussian kernel:

$$pdf(x) = \sum_{s=1}^{n} \frac{1}{nh\sqrt{2\pi}} e^{-\frac{(x-\Delta\theta_{y1,y2,s})^2}{2h^2}} \text{ with} \tag{4}$$

$h = 0.03$
$pdf(x)$: probability density function value over all grid points, [-]
$\Delta\theta_{y1,y2,s}$: projected change in blossom frost risk
x : any possible value of $\Delta\theta_{y1,y2,s}$, [-]
h : bandwidth of kernel smoothing window, [-]
s : grid point
n : number of grid points

$$pmf(x) = \frac{pdf(x)}{\sum_{j=1}^{z} pdf(j)}, \text{ with} \tag{5}$$

$pmf(x)$: probability mass function value over all grid points, [-]
z : number of possible values of $\Delta\theta_{y1,y2,s}$, [-]
j : index

Model description

Apple bloom was simulated using phenological models. In principle, models assume that the time of bloom is related to so-called temperature sums of chilling (Sc) and forcing (Sf), accumulated during winter (chilling phase) and spring (forcing phase) by the corresponding rates of chilling (Rc) and forcing (Rf). See tab. 1 for denominations.

$$Sc(t) = \sum_{i=t_0}^{t} Rc(T_i) \tag{6}$$

$$Sf(t) = \sum_{i=t_1}^{t_2} Rf(T_i) \tag{7}$$

Further it is assumed, that Sf is related to Sc as follows:

$$\text{Sequential models: } Sf(t_2) = a \cdot e^{bSc(t_1)} \quad (8)$$
$$\text{Parallel models: } Sf(t_2) = a \cdot e^{bSc(t_2)} \quad (9)$$

A basic thermal-time model (model 1) was applied as described, with the rate of forcing Rf:

Model 1

$$Rf(T_i) = \begin{cases} 0 & \text{if} \quad T_i \leq Tbf \\ T_i - Tbf & \text{else} \end{cases} \quad (10)$$

Sequential (model 2) and parallel (model 3) chilling-forcing models were applied as described in the following:

Models 2,3

$$Rc(T_i) = \begin{cases} 0 & \text{if} \quad T_i \leq 0 \text{ or } T_i \geq 10 \\ \frac{T_i}{Tbc} & \text{if} \quad 0 < T_i \leq Tbc \\ \frac{T_i - 10}{Tbc - 10} & \text{if} \quad Tbc < T_i < 10 \end{cases} \quad (11)$$

$$Rf(T_i) = \begin{cases} 0 & \text{if} \quad T_i \leq Tbf \\ \frac{28.4}{1+e^{(-0.185(T_i - Tbf - 18.4))}} & \text{else} \end{cases} \quad (12)$$

The Modified Utah model was applied for mean daily temperature values (model 4). Following a different approach, this model is a sequential model with Rc as in eq. 11 and with Rf being:

Model 4

$$Rf(T_i) = \begin{cases} 0 & \text{if} \quad T_i \leq Tbf \\ (T_i - Tbf) \cdot \left[1 + \left(\frac{Sf(T_{i-1})}{Sf(t_2)}\right)^2\right] & \text{else} \end{cases} \quad (13)$$

Due to findings for better performance when relating bloom additionally to radiation, models taking into account the length of the day were further included (models 5-7). Model 5 was applied in the version described, and being an extension of model 1 the rate of forcing is calculated as follows:

Model 5

$$Rf(T_i) = \begin{cases} 0 & \text{if} \quad T_i \leq Tbf \\ (T_i - Tbf) \cdot \left(\frac{D}{10}\right)^c & \text{else} \end{cases} \quad (14)$$

Models 6-7 are new variations of the sequential and parallel chilling-forcing models. These varied models also assume, that bloom is influenced by radiation only during the forcing phase. For both Rc was calculated as in eq. 11 and Rf was calculated as follows:

Model 6,7

$$Rf(T_i) = \begin{cases} 0 & \text{if} \quad T_i \leq Tbf \\ \frac{28.4}{1+e^{(-0.185(T_i - Tbf - 18.4))}} \cdot \left(\frac{D}{10}\right)^c & \text{else} \end{cases} \quad (15)$$

Table 1. Denomination of variables and parameters

Notation	Description	Unit
T	Air temperature	°C
Tbc, Tbf	Base temperature for chilling, forcing	°C
t	Time	hour [h], day [d] or year [a]
t_0	Start of the chilling period (dormancy)	day of the year (DOY)
t_1	Chilling requirement completed, start of forcing	day of the year (DOY)
t_2	Forcing completed (BBCH 60, BBCH 65)	day of the year (DOY)
Sc, Sf	State of chilling, state of forcing	—
Rc, Rf	Rate of chilling, rate of forcing	—
D	Daylength	h
a, b, c	Calibration parameters	—
i, s, z	Index variables	—
θ	Blossom frost risk	—
β	Temperature threshold for blossom frost	°C

Model parameters

Table 2. Model parameters (early ripeners, BBCH 65, area mean)

Model	Tbc [°C]	Sc [-]	Tbf [°C]	a [-]	b [-]	c [-]	t_1 [DOY]	t_2 [DOY]
1	—	—	5.8	—	—	—	—	122.6
2	3.0	36.9	5.0	220.9	-0.0248	—	12.1	120.4
3	2.5	37.8	3.1	201.3	-0.0029	—	17.4	121.9
4	4.2	37.7	7.4	—	—	—	17.4	121.9
5	—	—	0.7	—	—	1.3	[a]30.1	122.9
6	4.8	33.4	5.2	232.1	-0.0063	4.4	3.0	120.4
7	5.1	35.7	5.7	215.9	-0.0033	5.7	8.0	119.2

[a]This model does not calculate the fulfillment of dormancy, but optimizes t_1 as starting date for heat summation.

3.4 Future energy consumption of horticultural production in greenhouses

3.4.1 Objective

The objective is to project greenhouse energy consumption.

3.4.2 Summary

Energy costs are a main concern in greenhouse plant production. A decreasing energy demand for greenhouse heating could be expected in the mean, as temperatures in winter rise. However, mean, extent, uncertainty and regional influences are unknown. Therefore greenhouse energy demand was projected for IPCC emission scenarios A1B, A2 and B1 for the regions Germany and the state of Lower Saxony on different spatial resolution, using a validated energy simulation system. Additional computations with 2d bias corrected time series were carried out in a case study.

All results consistently show a mean decrease of energy demand. Hereby findings followed the projected warming closely, as major changes were projected beyond 2040. In spite of this trend, regional exceptions of single years have to be regarded. Furthermore, application of a 2d bias correction rendered similar results. The results show, how climate change will affect greenhouse infrastructure and production systems regarding the required heating system, greenhouse heating strategies and production strategies. Increasing energy costs might outreach energy savings and heating systems must probably withstand conditions similar to present until the mid-century. However, producing companies specialized on cultivation at different temperature set-points benefit unequally from energy savings. Hence, this might alter future greenhouse utilization concepts.

3.4.3 Publication: High Resolved Simulation of Climate Change Impact on Greenhouse Energy Consumption in Germany — *European Journal of Horticultural Science*

Authors:	Holger Hoffmann[1] and Thomas Rath[2]
Journal:	European Journal of Horticultural Science
Volume:	77(6)
Page:	241-248
DOI:	—
ISSN:	1611-4426 (print version), 1611-4434 (electronic version)
Publisher:	Eugen Ulmer KG
Address:	Stuttgart, Germany
Date of Submission:	12 July 2012
Date of Acceptance:	08 October 2012
Date of Publication:	03 December 2012
Current Status:	Published

Contribution of authors:

1: Data acquisition and processing, calculations, evaluation, manuscript development and processing

2: Review

This publication can be accessed via `http://www.ejhs.de/`
and is indexed by 'scopus' and 'web of knowledge' among other search engines.

High Resolved Simulation of Climate Change Impact on Greenhouse Energy Consumption in Germany

H. Hoffmann and T. Rath
(Institute of Biological Production Systems, Biosystems and Horticultural Engineering Section, Leibniz Universität Hannover, Hannover, Germany)

Summary

Increasing energy costs are a main concern in greenhouse plant production. At this it is uncertain, how the expected climate change will affect this problem regionally. Therefore the future greenhouse energy consumption was simulated with the help of climate realizations of high spatio-temporal resolution from the regional climate model REMO (UBA-Runs), based on the IPCC climate projections A1B, A2 and B1. Simulations were conducted for each hour of the periods 2001–2015 and 2031–45 on a 100 km × 100 km grid for Germany, as well as continuously for each hour from 1951 to 2099 on a 10 km × 10 km grid for the federal department Lower Saxony (Germany), employing the energy simulation system HORTEX. Furthermore the influence of a consistent 2-d bias correction on the projected signal was tested in a case study (53° 03' N, 08° 48' E). The results consistently show a strong mean decrease of greenhouse energy consumption for all scenarios by 2038 and up to 45 % for the area of Germany, diverging regionally. Higher absolute reductions in energy consumption can be expected in warm greenhouses, while low temperature set points result in higher relative energy consumption reduction. The latter might influence future utilisation concepts.

Key words. bias correction – climate change – energy consumption – greenhouse

Introduction

The increase of energy costs of the last decades to a present crude oil price of above one hundred dollars per barrel (USEIA 2012), as well as the influences of the increasing scarcity of resources or environmental standards along with pricing policy are indicators of a lasting trend. Being potentially existence-threatening, this is a main concern to energy-intensive sectors. Since the energy consumption of German greenhouse production is up to 0.5 % (calculation based on data from STATISTISCHES BUNDESAMT 2006, 2012) of the total industrial energy, it is important to analyse possible changes of greenhouse energy use in relation to climatic change effects. Besides fluctuating energy costs it is further unknown, how climate change will affect the energy consumption of greenhouses regionally in the long run. At this, changes in driving forces as temperature or radiation will influence future energy consumption. While climatic warming can be expected to reduce the energy demand for greenhouse heating in the mean, particular attention should be paid to the utilisation concepts of greenhouses regarding cold and warm temperature strategies, as well as to the regionally varying climatic impact. Despite HOFFMANN and RATH (2009) investigated the climatic impact on greenhouse energy consumption, no literature has been published so far concerning simulation studies with different spatial resolution and optimization of the input data.

Objectives

Highly resolved simulations of future greenhouse energy consumption are necessary to consider regional climatic changes. For this uncorrected and locally bias corrected climate data should be consequently applied and interpreted in case studies to simulate the pattern of energy consumption.

Material and Methods

Simulation input: Greenhouses

In order to estimate the impact of the regional climate change, future greenhouse energy consumption was simulated with the help of the greenhouse energy simulation system HORTEX (RATH 1994, 2006). Three different simulations were conducted with year-round lower temperature set point values (day/night) of 5/5 °C and 18/16 °C. Greenhouse settings for energy simulation are shown by Table 1.

Table 1. Basic simulation input settings for HORTEX.

Domain	Input parameter	Unit	Setting
Climate	hourly ambient air temperature[a]	°C	varying sim. parameter
	hourly global radiation	W m^{-2} s^{-1}	varying sim. parameter
	average wind speed	m s^{-1}	4
Greenhouse	construction	–	Venlo-type
	U'-value	W m^{-2} K^{-1}	7.6
	ground area	m^2	10000
	greenhouse cover	–	single glazing
	energy screen	–	one-layer standard
	side wall height	m	4
Control	heating set point (day / night)	°C	5/5; 18/16
	venting set point	°C	30
	energy screen threshold	W m^{-2} s^{-1}	0.1[b]
	assimilation lighting	–	none
	heating system	–	mixed

[a] in 2 m height above ground
[b] global radiation

Simulation input: Climatic Data

Regional projections based on simulated climate time series were conducted to investigate the impact of a climate change on greenhouse energy consumption. Climate data from realizations (first run) from of the regional climate model REMO (driven by the general circulation model ECHAM5) of the Intergovernmental Panel on Climate Change (IPCC) climate projections A2, B1 (simulation 1) and A1B (simulations 2 and 3) were used (MPI 2012). Hereby the IPCC-scenarios, resulting in different greenhouse gas emission scenarios (SRES) on which climate projections are based, assume either a heterogeneous world with continuously increasing population and regionally oriented economic development (A2), or a convergent world with a population that peaks in mid-century and declines thereafter as well as fast changes in economic structures (B1, A1B). Hence the present findings follow the chain of information: IPCC scenario > emission scenario > climate projection > climatic impact simulation.

According to these scenarios, air temperature in Central Europe will rise by 2.5 °C (B1) and 3.7 °C (A1B, A2) by the end of the century as compared to 1961–1990 (ROECKNER et al. 2006). Hereby increases of temperature will mainly occur during the 2nd half of the century with temperature increases for Germany during the period investigated (simulation 1) of 0.1 °C (B1) and 0.9 °C (A2) by 2038 as compared to 2008 (both 15-year- mean).

For the simulation the air temperature (in 2 m height above ground) was corrected for altitude (−0.0064 °C m^{-1}).

Global radiation was calculated from the net down- and upward surface radiation of the REMO data (MPI 2012). Wind speed was set constant to 4 m s^{-1}, both in order to account for regional/small scale effects and to ensure the comparability of the results.

Simulation 1: Greenhouse energy consumption in Germany

Simulated hourly climate data from the periods 2001–2015 and 2031–2045 of approximately 10 km (0.088°) resolution was aggregated to 57 areas in Germany of approximately 100 km × 100 km by taking the mean of all grid points inside the corresponding area. Subsequently, the mean of each of the 8760 hours of the year was calculated for each period (see also HOFFMANN and RATH 2009).

Simulation 2: Greenhouse energy consumption in Lower Saxony

A second simulation was conducted for each hour from 1951 to 2099 on higher resolution for the area of Lower Saxony (Germany). The spatial mean of areas consisting of 3 × 3 grid points (~30 km × 30 km) was calculated for 783 adjacent grid points, hence obtaining the spatial floating mean of 783 overlapping areas with a grid resolution of 10 km × 10 km. Unlike simulation 1, simulation 2 was conducted using each hour of each year of the climatic data instead of taking the mean of each hour over the years. Final results of energy consumption were brought to a regular grid by ordinary kriging (WACKERNAGEL 1995) for the purpose of visualization.

Simulation 3: Greenhouse energy consumption simulated with bias corrected climatic data (case study Bremen)

In order to estimate the influence of the simulated climate data quality on the computed energy consumption, the latter was additionally computed with bias corrected climate data in a case study. For this, hourly climate data (1951 to 2099) at 53° 03' N, 08° 48' E from the German Meteorological Survey (DWD) was used to remove the bias of the simulated data by applying distribution based bias correction (Quantile mapping), established by INES and HANSEN (2006) and PIANI et al. (2010). To restore the necessary climate variable consistency, quantile mapping was applied 2-dimensionally as described by HOFFMANN and RATH (2012). For this, correction or transfer functions from simulated to measured data were obtained by mapping air temperature and global radiation non-parametrically by applying a gaussian kernel with bandwidth h = 0.1 (BOWMAN and AZZALINI 1997) and a optimization factor K = 0.5 (see HOFFMANN and RATH 2012). Hereby transfer functions were derived from 1977-2010 and applied further to REMO data from 1951-2099.

Results

According to simulation 1, greenhouse energy consumption will decrease in the mean, regardless of temperature setting or scenario. For Germany, area average reductions up to 45 % and up to 45 kWh m^{-2} year^{-1} were found, depending on greenhouse temperature settings and on the chosen climate scenario (Fig. 1, 2). At this, stronger decreases were found in the warmer scenario A2 than in scenario B1. Further, stronger absolute but lower relative reductions (average 10 %) were found for higher greenhouse temperature settings of 18/16 °C (Fig. 2), while settings of 5/5 °C led to area average reductions of up to 50 % (Fig. 1). Unlike these mean tendencies, regional variation ranges from reductions of 89.6 % to an increase of 3.5 %, although no increases were found for the warmer scenario A2. Hereby low temperature settings led to a stronger decrease in south-east of Germany (which exhibits higher altitude) than in the north. This pattern was also found for higher temperature settings within in the scenario B1, but not for scenario A2. The latter shows an 'east-to-west distribution' with higher reductions in the east.

Simulation 2 for Lower Saxony confirmed the findings from scenario A2 on a 100 times finer grid for scenario A1B (Fig. 3, 4). Hereby energy consumption reflects the orography of the region, e.g. with the higher elevated south-east of Lower Saxony displaying higher energy consumption. Past to present energy consumption is apparently dominated by a periodicity of cold and warm climate cycles (Fig. 3, 4), fluctuating around the present level. Nevertheless the downward trend in energy consumption arises by the mid of the 21st century, manifesting further in the distribution of the annual sum of greenhouse energy consumption (Fig. 5). As shown, a stronger mean shift of the distribution is found from 2011-2040 to 2061-2090, than from 1961-1990 to 2011-2040. Furthermore, regarding the extremes of the distribution, a shift of the mean is

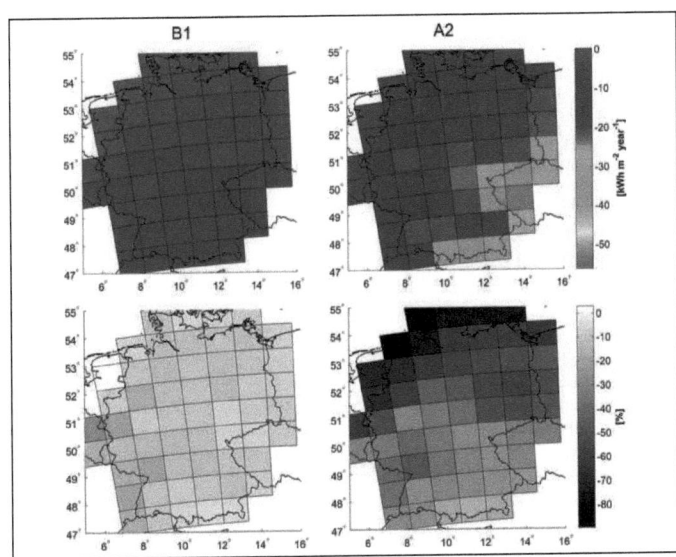

Fig. 1. Projected changes in greenhouse energy consumption in Germany by 2031–2045 as compared to 2001–2015 for scenarios B1 and A2 and temperature set-points 5/5 °C.

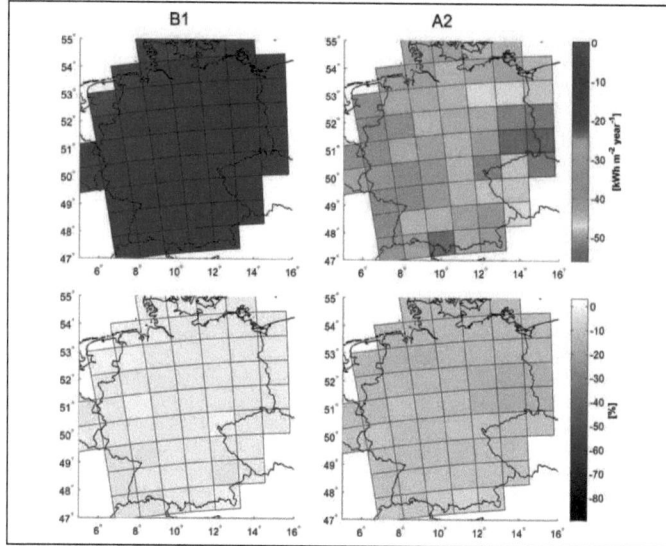

Fig. 2. Projected changes in greenhouse energy consumption in Germany by 2031–2045 as compared to 2001–2015 for scenarios B1 and A2 and temperature set-points 18/16 °C.

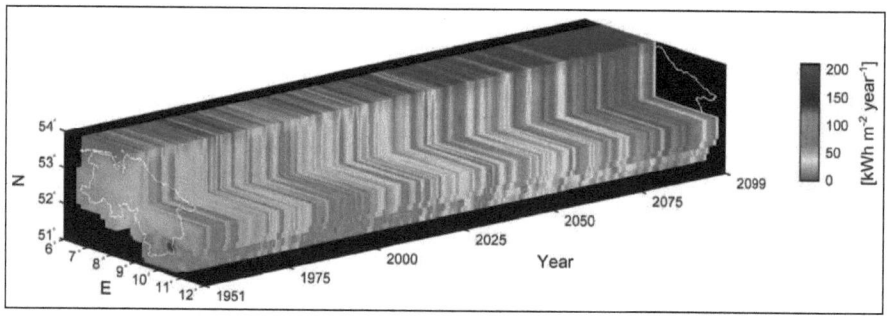

Fig. 3. Projected greenhouse energy consumption in Lower Saxony simulated for the climate scenario A1B and temperature set-points 5/5 °C (day/night).

observed prior to the reduction of the upper extreme, resulting in an increased variance.

Regarding simulation 3, a bias correction of the simulated climate data (REMO) for Bremen decreased the mean deviation from simulated to measured climate variables (Table 2). Hereby underestimation of global radiation and overestimation of mean air temperature were removed, resulting in an overestimation of global radiation by ~1 % (1977–2010). However, reduction of the previous climate model bias resulted in discrepancies of the energy consumption calculated from measured and simulated-corrected climate data (~3 %). At this, the range of energy consumption in terms of annual sums from measured data was fairly reproduced with both simulated and simulated-corrected climate data (see grey rectangles, Fig. 6). Years with elevated energy consumption exhibited further increased energy consumption after bias correction of the underlying data. While energy consumption from corrected data was underestimated in the period from 1977–2010 compared to calculations from measured data, projection for all REMO years with corrected climate data resulted in higher energy consumption compared to the projection with uncorrected climate data (Table 2, Fig. 6). Hereby energy consumption was in-

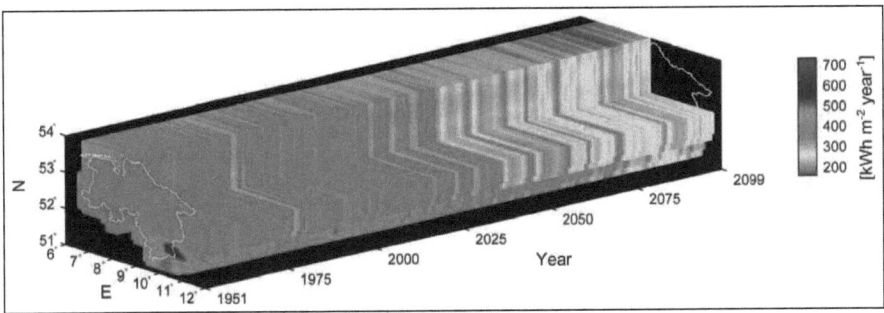

Fig. 4. Projected greenhouse energy consumption in Lower Saxony simulated for the climate scenario A1B and temperature set-points 18/16 °C (day/night).

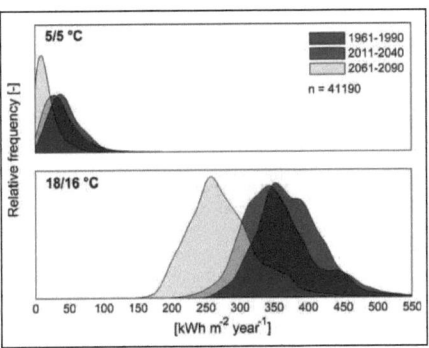

Fig. 5. Distribution of the projected yearly greenhouse energy consumption in Lower Saxony simulated for day/night temperature set-points of 5/5 °C and 18/16 °C and climate scenario A1B. The distribution was estimated by applying a smoothing gaussian kernel with bandwidth h = 8.

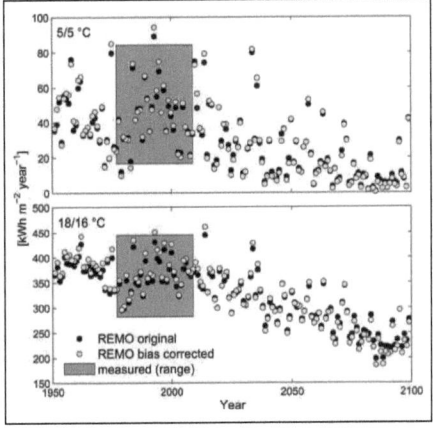

Fig. 6. Projected yearly greenhouse energy consumption for different day/night temperature set-points, calculated from original and bias corrected simulated climate data in Bremen (scenario A1B). Rectangles indicate the period and range (min, max) of energy consumption computed from measured values.

Table 2. Influence of the bias correction on climate data quality and simulated energy consumption (1977–2010).

Climate Data	Global radiation[a] [kWh m^{-2} year^{-1}]	Air temperature[b] [°C]	Energy consumption[a] [kWh m^{-2} year^{-1}]
Measured	960.8 ± 76.2	9.1 ± 1.3	354.3 ± 44.5
Simulated (original)	897.2 ± 76.0	9.4 ± 1.0	354.8 ± 36.0
Simulated (bias corr.)	970.9 ± 67.3	9.1 ± 1.0	343.2 ± 38.7

[a] mean and standard deviation of annual sums
[b] mean and standard deviation of annual means

creased in the mean by 0.1 and 2.2 kWh m^{-2} year^{-1} for temperature settings of 5/5 and 18/16 °C, respectively.

Discussion

Influences of air temperature, greenhouse temperature and altitude

In detail the magnitude of expected reduction in energy consumption mainly depends on the three following factors (i) climatic data as depending on the chosen scenario and resolution, (ii) greenhouse temperature settings and (iii) orography of the domain investigated. As the projected future warming of the scenario B1 is lesser than of the scenarios A2 and A1B (ROECKNER et al. 2006), future energy consumption for scenario B1 does not decrease as far as for scenarios A2 or A1B. Further, energy consumption is influenced by the site orography, as air temperature decreases with increasing altitude, leading to higher energy consumption at higher altitudes. As this influence remains constant for the different climate scenarios, it can be considered as a function of spatial resolution of the climate data or simulation. The present simulations resolved the elevated energy consumption of the low mountain range (~200–500 m altitude) on a 10 km × 10 km grid for Lower Saxony, whereas these findings could hardly be distinguished through a resolution of 100 km × 100 km. Furthermore these results reflect the stronger temperature increase in higher elevated areas, the higher elevated south-east of Lower Saxony displays the sharpest decline in greenhouse energy consumption.

Influence of climate data bias correction

Depending on the climate model, climate variable, timescale, measurements and gridding (resolution) among others, simulated climate time series may have large biases to measured data (HAERTER et al. 2011, MARAUN 2012). The removal of this bias is possible (bias correction), e.g. by means of quantile mapping (PIANI et al. 2010). Nevertheless 1-dimensional bias correction of single climate variables separately can lead to large errors in multidimensional impact models (MARAUN et al. 2010). Therefore a consistent bias correction approach (HOFFMANN and RATH 2012) was used in the present study to restore climate variable consistency. In this work projected energy consumption with uncorrected and corrected simulated climate data differed in the mean by 0.3 % and 0.7 % for temperature settings of 5/5 °C and 18/16 °C respectively (case study Bremen, 1951-2099). Hence, the climate impact signal was hardly influenced by the bias correction. Nevertheless, simulated annual energy consumption with bias corrected climate data exhibited larger variation than without bias corrections, reproducing the variance simulated from measured data slightly better (1977–2010). However, bias correction of climate variables (1977–2010) improved climate variable quality (bias, consistency) but led to slight underestimation of the resulting energy consumption. Despite this discrepancy, projected energy consumption with bias corrected data are possibly more robust due to the following: i) Uncorrected simulated climate data overestimated temperature and underestimated global radiation, possibly resulting in energy consumption similar from that calculated from measured climate data. Hence, the true absolute error of the energy simulation with uncorrected climate data would be larger than from measurements, but did sum up to a smaller bias. ii) The intra-annual course of energy consumption from uncorrected data exhibits the characteristic patterns inherited from the regional climate model, being different to the pattern of the intra-annual energy consumption from measured climate data.

Influence of concatenating climate and impact models

The stated energy simulations use climate simulations (MPI 2012) based on emission scenarios which describe possible future developments of the world (IPCC SRES 2000). Hereby general circulation models generate climatic data, which are downscaled to regional resolution via regional climate models. To decrease deviation and inconsistencies in the data, bias correction procedures are applied. At the end impact models use these corrected data. Since this concatenation is a source of large uncertainties in projections, simulation uncertainty increases with increasing time horizon of the projection (HAWKINS and SUTTON 2009).

Timeline of the climatic impact on energy consumption

According to the present results, major changes in greenhouse energy consumption are not to occur before 2035. While mean reductions in energy consumption can be expected from the mid of the century on, year to year differences (variance) will increase. These findings are consistent with the stronger climatic impact in the second half of the century. Therefore, with single years until the mid of the century potentially demanding energy consumption at the present level, changes in heating strategies can be expected only concerning utilization concepts of greenhouses. Consequently changes of the employed heating systems are therefore unlikely to occur before 2050.

Conclusions

The present findings can be summarized: 1.) Climatic warming can be expected to lead to a mean reduction in greenhouse energy consumption of 10 % and 50 % for high and low temperature settings respectively. 2.) Reduction is regionally highly divergent as being dependent on site orography. 3.) Greater changes of the mean level of energy consumption cannot be expected before the mid of the century. 4.) Despite a mean reduction of energy consumption single year extremes potentially reach the present level till the mid of the century. 5.) Changes in energy consumption could lead to innovations in greenhouse utilization concepts for low temperature strategies (AKYAZI and TANTAU 2012). 6) Consistently bias corrected time series of dynamical-physical climate models like Remo or CLM should be used where possible. If bias correction is not possible, original data of the climate model must be used with care, since simulation results might depend on the systematic error of the climatic input. Alternatively statistical regional climate models as WETTREG (KREIENKAMP 2006) could be employed.

Looking at the specific results for German greenhouse energy consumption it can be concluded that, since operation at higher temperatures usually leads to a higher portion of energy costs (which are expected to increase for the future) compared to total variable costs, innovation of greenhouse utilization concepts are more likely to occur for lower temperature settings. Therefore potential benefits of the climatic warming due to absolute energy consumption reduction in warm-house cultivation (≥ 18 °C) as for ornamental plant production, e.g. orchids, or vegetable production, e.g. cucumber or tomatoes, (KRUG et al. 2007) can be expected to be diminished through increasing energy costs.

Acknowledgements

The study was supported by the Ministry for Science and Culture of Lower Saxony within the network KLIFF – climate impact and adaptation research in Lower Saxony.

References

AKYAZI, G. and H.-J. TANTAU 2012: ZINEG – The low energy greenhouse. An innovative greenhouse with new climate control strategies supported by phytomonitoring data. In: CASTILLA, N., O. VAN KOOTEN, S. SASE, J.F. MENESES, W.H. SCHNITZLER and E. VAN OS (eds.): XXVIII International Horticultural Congress on Science and Horticulture for People (IHC2010): International Symposium on Greenhouse 2010 and Soilless Cultivation. ISHS Acta Hortic. 927. ISBN 978-90-66057-24-1.
BOWMAN, A. W. and A. AZZALINI 1997: Applied Smoothing Techniques for Data Analysis. Oxford University Press, New York.
HAERTER, J. O., S. HAGEMANN, C., MOSELEY and C. PIANI 2011: Climate model bias correction and the role of timescales. Hydrol. Earth Syst. Sc. **15**, 1065–1079.
HAWKINS, E. and R. SUTTON 2009: The potential to narrow uncertainty in regional climate predictions. B. Am. Meteorol. Soc. **90**, 1095–1107.
HOFFMANN, H. and T. RATH 2009: Überregionale Simulationen zum zukünftigen Energieverbrauch von Gewächshäusern unter Berücksichtigung von IPCC-Szenarien. In: Anforderungen an die Agrarinformatik durch Globalisierung und Klimaveränderung. ISBN 978-3-88579-236-9. GIL **21**, 61–64.
HOFFMANN, H. and T. RATH 2012: Meteorologically consistent bias correction of climate time series for agricultural models. Theor. Appl. climatol. (accepted 22.2.2012). doi: 10.1007/s00704-012-0618-x.
INES, A.V.M. and J.W. HANSEN 2006: Bias correction of daily GCM rainfall for crop simulation studies. Agr. Forest Meteorol. **138**, 44–53.
IPCC SRES 2000: NAKIĆENOVIĆ, N. and R. SWART (ed.): Special Report on Emissions Scenarios: A special report of Working Group III of the Intergovernmental Panel on Climate Change, Cambridge University Press, ISBN 0-521-80081-1, 978-052180081-5 (pb: 0-521-80493-0, 978-052180493-6).
KREIENKAMP, E. 2006: WETTREG A1B SCENARIO RUN, UBA PROJECT, 2001–2010. World Data Center for Climate. CERA-DB „WR_A1B_2001_2010". URL: http://cera-www.dkrz.de/WDCC/ui/Compact.jsp?acronym=WR_A1B_2001_2010 (last fetch: 23.06.2012).
KRUG, H., A. ROMEY and T. RATH 2007: Decision support for climate dependent greenhouse production planning and climate control by modelling. II. Modelling plant growth. Eur. J. Hortic. Sci. **72**, 145–151.
MARAUN, D., F. WETTERHALL, A.M. IRESON, R.E. CHANDLER, E.J. KENDON, M. WIDMANN, S. BRIENEN, H.W. RUST, T. SAUTER, M. THEMEL, V.K.C. VENEMA, K.P. CHUN, C.M. GOODESS, R.G. JONES, C. ONOF, M. VRAC and I. THIELE-EICH 2010: Precipitation downscaling under climate change: recent developments to bridge the gap between dynamical models and the end user. Rev. Geophys. **48**, RG3003.
MARAUN, D. 2012: Nonstationarities of regional climate model biases in european seasonal mean temperature and precipitation sums. Geophys. Res. Lett. **39**(6).
MPI 2012: Max Planck Institute for Meteorology, Hamburg. Regional Climate Modelling – REMO. URL: http://www.remo-rcm.de/REMO-UBA.1189.0.html (last fetch: 26.04.2012).
PIANI, C., J. HAERTER and E. COPPOLA 2010: Statistical bias correction for daily precipitation in regional climate models over Europe. Theor. Appl. Climatol. **99**, 187–192.
RATH, T. 1994: Influence of thermal storage on the calculation of greenhouse heat consumption with k'(U)-value-models. Gartenbauwissenschaft **59**, 39–44.
RATH, T. 2006: Hortex 3.0 und Hortexlight 1.0- grafisches Softwaresystem zur Planung der Energieversorgung von Gewächshausanlagen. URL: http://www.bgt.uni-hannover.de/software/ (last fetch: 26.04.2012).
ROECKNER, E., G.P. BRASSEUR, M. GIORGETTA, D. JACOB, J. JUNGCLAUS, C. REICK and J. SILLMANN 2006: Climate projections for the 21st century. Max Planck Institute for Meteorology, Hamburg, Germany.
STATISTISCHES BUNDESAMT 2006: Gartenbauerhebung 2005, Fachserie 3, Artikelnummer 2032001059004.

STATISTISCHES BUNDESAMT 2012: Energieverwendung in der Industrie. URL: https://www.destatis.de/DE/ZahlenFakten/Wirtschaftsbereiche/Energie/Verwendung/Tabellen/VerwendungIndustrie.html. Last fetch: 05.10.2012.

USEIA 2012: U.S. Energy Information Administration: URL: http://www.eia.gov, (last fetch: 26.06.2012).

WACKERNAGEL, H. 1995: Multivariate Geostatistics – An Introduction with Applications, Springer Berlin, 79 ff.

Received 07/12/2012 / Accepted 10/08/2012

Addresses of authors: Holger Hoffmann (corresponding author) and Thomas Rath, Institute of Biological Production Systems, Biosystems and Horticultural Engineering Section, Leibniz Universität Hannover, Herrenhäuser Str. 2, 30419 Hannover, Germany, e-mail: hoffmann@bgt.uni-hannover.de

Chapter 4

Closing remarks

4.1 General remarks on presented investigations

The presented methods and simulations were partially carried out for areas as well as single locations across Germany and in higher resolution and detail for the federal department Lower Saxony (Germany). Accordingly, the results are confined to a regional climate change at first glance. Therefore a broader discussion is made in the following, pursuing generalized conclusions. The methodology is finally put into a common framework, going from climate data processing, followed by plant reactions, to horticultural production systems.

4.2 Résumé of specific climate change effects on horticultural production

At first, each work is regarded as a case study. With climate projections heavily relying on simulated climate data, the need for a consistent bias correction of climate variables was given. While no bias correction is able to maintain physical consistency among climate variables so far, the developed method improved bias as well as consistency. The latter was deduced by comparison to probabilities of observed meteorological states. Thus, the method operates purely statistically and is furthermore evaluated statistically. The same idea was followed by (Piani and Haerter 2012) who presented a similar approach, thus emphasizing the necessity for improved bias correction. This second approach is straightforward, applying "Quantile mapping" on one climate variable for segments of a given quantile range from a corresponding second climate variable. While both approaches were tested with hourly data of point measurements in two dimensions, independent projections of climate or climate impact with these approaches have not been published so far. Nevertheless one simulation was carried out for greenhouse energy demand (see chapter 3.4.2), extending the approach to the 2d combination of global radiation and temperature. Hence, albeit validation of methods, uncertainty assessment for ensembles projections of impact studies is yet to come. Furthermore, so far only the 2d cases of precipitation and global radiation (Hoffmann and Rath 2012b), temperature and global radiation (Hoffmann and Rath 2012a) and temperature and precipitation (Piani and Haerter 2012) are known. It would be intriguing to verify results of higher dimensional cases or cases with different underlying 1d bias corrections or of climate variables with different distributions than normal or gamma-distributions.

Water deficiency determines plant production as shown for leafy vegetables by the example of lettuce (see 3.2.2). Thereby a basic model was developed in order to be able to project the future impact of changes in precipitation on vegetable production. Based on radiation and temperature driven growth, stress in the form of growth reduction is simulated in proportion to soil moisture as depending on water supply and evapotranspiration, with the latter again depending on plant size. The model can therefore be taken as a general model for most annual/herbaceous crops. Simulation for other species could be conducted after parametrization and replacement of plant diameter by leaf area index (LAI) if necessary. Due to the general character of the model, an application with simulated time series seems possible. This would however require a deeper knowledge of interactions of CO_2 and water use efficiency, as trends in projected yield depend largely on these (see 1.3.2). Therefore future precipitation patterns were not projected in the present work with regard to impact on plant production. Hereby precipitation patterns diverge regionally, with observed and projected decreasing summer precipitation in Lower Saxony (see 1.2.9). While this increases the irrigation water demand, shorter crop cycles can be expected for crops similar to lettuce due to rising temperature. Nevertheless, drier summers have not led to significant changes in yield since 1991 (from LSKN 2013) as irrigation is common in horticultural production. Lettuce itself is however a cool-season crop, grown mainly in spring and late summer. Hereby total precipitation in spring has increased during the past (Haberlandt et al. 2010) and is project to increase for the future (Moseley et al. 2012). Furthermore, future decreasing summer precipitation cannot be expected to lead to significant changes of yield in late summer grown lettuce, as irrigation takes place and water allocation for horticultural crops is not expected to decrease (Rubino et al. 2012). Additionally, no changes were projected for the length of dry periods from April to September (Moseley et al. 2012). Hence, albeit effects of changing CO_2 as well as whole year water balance were not taken into account in the present work and precipitation frequency and pattern are not represented by mean precipitation, decreasing summer precipitation can be considered to be of low risk. However, due to the large uncertainties of modeling precipitation, final statements for the risk for horticultural crops through changes in precipitation cannot be given and must further cover the mentioned aspects.

Rising temperatures are likely to impact horticultural production at all scales. Hereby trends observed during the past (see 1.3.4) are not necessarily going to continue for the future and new trends might become apparent. These trends are based on temperature effects, which can be separated into a direct impact on phenology/development (e.g. on vernalization), a direct impact on growth related processes (e.g. photosynthesis) as well as indirect impacts on both (e.g. through soil moisture). As a results, all effects may challenge crop production by altering cultivation and/or yield. The present work comprises all effects, though focusing on chilling and blooming (development). Temperature dependent growth was described for lettuce (see above) and indirect effects were marginally incorporated by taking evapotranspiration into account for lettuce growth. Effects on phenological phases which are directly influenced by human activity were not investigated (so-called "false" phases, Menzel and Sparks 2006: 78, e.g. dates of harvest).

Phenological timing of fruit trees and annual plants has changed during the past, mainly observed through earlier bud-burst and flowering observed with trees (see 1.3.4). Albeit a slowdown of these processes has not been reported, a constant trend for plants with narrow temperature optima or fixed temperature thresholds regarding temperature requirements for phenological development seems unlikely for the future. As shown for apple blossom, an ongoing advancing of flowering is projected at a lower pace than in the past. This stands in line with theory (Körner 2006: 64; Menzel and Sparks 2006: 85Legave et al. 2008a) as delayed chilling fulfillment counters the faster spring forcing phase. The difference to results from simple Thermal Time (Heat Sum) models (Chmielewski et al. 2005; Hoffmann et al. 2012), which do not account for a chilling period, is shown thereby. These differences of pure forcing and chilling-forcing models give an insight on the magnitude of the chilling and following consistent effects. On the one hand, years with unfulfillable chilling requirement cannot be ruled out for the future. The use of chemicals for breaking of dormancy as applied already today in lower latitudes would be a consequence. However, this signal was not significant and needs further analysis, as the low appearance of such years requires extreme event analysis. On the other hand, last spring freeze and onset of flowering are not projected to advance in parallel, eventually resulting in a decreased blossom frost risk. Hereby changes in projected blossom frost risk were exceeded by internal and model variability. However, this is in agreement with general findings for tree flowering and shifts in spring freeze (Scheifinger et al. 2003; Menzel and Sparks 2006: 93,Moseley et al. 2012), but contrasts other studies of blossom frost stating increasing blossom frost risk. The latter cannot be used for direct comparison due to discrepancies in region and modeling approaches (see 3.3.2), largely relying on single climate realizations, simple Thermal Time models and not presenting statistics. Furthermore, earlier flowering species were affected stronger than late flowering species, coinciding with the relative flowering trends of these two groups (Abu-Asab et al. 2001). However, none of the projections account for future changes in planted varieties, hence ignoring e.g. the range of plants exhibiting a lower chilling requirement. Varieties could also benefit differently from changing blossom frost as they differ in their susceptibility (Rugienius et al. 2009). While the climatological suitability of species and varieties will shift geographically, the composition of regional varieties produced today are likely to change. Hence, changes in varieties might alter the mentioned effects from decreasing chill availability, possibly also buffering the mentioned risks.

A second critical aspect regarding plant phenology and climate change is obligate vernalization. Changes in temperature away from the optimal vernalization temperature will lead to delayed or unfulfilled vernalization. Thereby, this can no longer be regarded as a "Knock-Out-effect" for Central Europe, as the projected temperature increases will always be outreached in the course of a year. However, devernalization through increasing frequency of higher temperatures may take place in some species, e.g. head cabbage (Krug et al. 2002: 293). These effects may be positive or negative, depending on the plant organ produced for sale. Furthermore, a delay in vernalization could be compensated by faster completion of a juvenile and/or curd growth phase. Although not shown by the present work, this would however lead to changes in the production schedule or chosen variety. Again, early to late varieties differ in their requirements for vernalization as well as temperature response concerning growth and were projected to respond differently to rising temperatures (Hoffmann and Rath 2013a,b). Hereby early variety cultivation time hardly differed from the present level(+0.4 to +2 d), whereas mid to late varieties showed a decreasing cultivation period (-18.1 to -3.7 d). This includes the variance of production duration for planting dates of a given month.

Finally, heat stress due to increasing frequency and extent of extreme temperatures is likely to increase, as the distribution of temperatures shifts to to higher values (see theoretical assumption about distribution, fig. 1.5). Hereby these effects might show a stronger impact than changes in precipitation, as Semenov and Shewry (2011) found larger reduction of wheat yield (simulation for 2050) by heat stress at flowering than for drought, due to earlier maturing and hence avoiding severe summer droughts. So far, heat stress for horticultural open field crops has not been projected. However, greenhouse crops might experience heat stress at higher frequency (Boulard et al. 2011). The latter will certainly limit the operation range of closed cultivation system.

Future climatic change impacts on horticultural production were considered in the present work by projection of apple blossom as discussed above, as well as by projection of future greenhouse energy consumption. A mean decrease of the latter will affect cultivation at high and low temperatures differently, as production at higher temperatures usually exhibits a higher portion of costs of the total variable costs. Hence, despite an expected general increase of energy costs due to rising energy prizes, larger savings will occur at higher temperatures. Nevertheless, while profit strongly depends on the market, rising temperatures are likely to increase the range of possible greenhouse utilization concepts at lower temperatures. As increasing air temperatures lead to a large relative decrease in energy demand of up to 90 % at lower temperatures, the range of possible/profitable crops could be extended.

4.3 Résumé of general climate change effects on horticultural production

4.3.1 Projection framework

Applying high-resolution climate time series in order to project the long-term climatic impact may seem paradox. Hereby the applied methodology (data processing, bias correction, impact projection) is commonly used to project the climate impact for the 21st century. Using this process, long-term changes are captured by transferring physical-dynamical projections of climate into purely statistical information on climate impact. The statistical signal emerges as climate data is interpolated and bias corrected as well as by considering long-term trends and variation of the climatic impact across the time series itself, different models and finally scenarios. Hereby highly resolved time series are either required for impact models (e.g. HORTEX) or in order to increase the information on spatial variance. Consequently, differences in simulated and measured climate data are larger than at coarser resolution and hence often require a correction for the use with impact models. With this approach, climate model improvement cannot be expected to change this situation in the short run, but it would be intriguing to know, when RCM performance will perform on a straight useful level for impact models, as the remaining bias is largely inherited by the driving GCM. Furthermore, as long as bias correction will be required, multidimensional approaches are necessary in order to apply impact models with more than one climate variable as input.

The presented methodology adopts the scenario approach of the IPCC (see 1.2.1). Hence, uncertainties of scenario as well as of climate projections have to be taken into account. Hereby the range of the former can be accounted for by projecting climate impact for all possible futures (all scenarios). The present work employed the scenarios B1, A1B and A2 with focus on scenario A1B. While all scenarios are meant as equally likely possible alternatives of the future, all scenarios show trends of a warming world with best estimates of temperature increases from 0.6 °C (scenario "Constant Year 2000 concentrations") up to 4.0 °C (scenario A1FI) (Solomon et al. 2007: 13, 2090-2099 minus 1980-1999). However, a scenario with "Constant Year 2000 concentrations" may be useful only for reasons of comparison, as recent trends indicate ongoing increasing emission of greenhouse gases. Without taking this scenario into account, the projected lowest increase in global surface temperature is projected with 1.8 °C (B1, best estimate). The scenario A1B can thus be regarded as an average scenario (best estimate: +2.8 °C). Hereby, scenarios A1B and A2 show similar trends for Lower Saxony. Thus, the presented projections can be regarded as an indicator for moderate or mean climate change, noticing large differences to other scenarios as shown for greenhouse energy consumption and scenario B1 (see 3.4.2).

Climate projection uncertainty directly affects the estimated climatic impact and the reliability of climate models is still under debate (van Oldenborgh et al. 2013), though given their usefulness (see 1.2.8). Hereby, the resulting uncertainty for the projected climate impact mainly depends on climate and impact model properties, which are assessed through using several model runs (climate variability) as well as several models (model properties). These uncertainties were taken into account by variance decomposition of the climate impact signal in combination with single time series statistics. Putting them into relation to the variance across climate scenarios, which are usually larger, their conceptual differences have to be taken into account: The variance across scenarios shows the broad range of possibilities across alternatives futures, resulting from projections for a chaotic system comprising human and natural activity. The variance across models shows the detailed assessment of one possible future, resulting from projections for a physical environmental system comprising climate and model properties. Here, the presented methodology was used to assess the projection uncertainty for a given future, thereby separating the effects of climate model and impact model variance. Summarizing, the approach shows the precision reached for a scenario ensemble member while giving an estimate for the scenario ensemble mean (A1B) and loose estimates for the scenario ensemble variability (B1, A2).

4.3.2 Future trends and risks in horticultural production

Given the projected mean changes of the present work, the following risks and tendencies for future regional horticultural production can be summarized (tab.4.1), complementing or modifying estimates given in section 1.3.5. Strong risks or Knock-Out-Effects were not detected as shown. Therefore, no significant yield depressions are expected. This does however not exclude changes in timing and length of cultivation as discussed above (see 4.2). Finally, a generally negative long-term impact of climate change, as stated for German agriculture (Zebisch et al. 2005: 75), cannot be affirmed for regional horticultural production.

Table 4.1: Trends and future risks of abiotic impact of climate change for selected horticultural aspects

Parameter	Effect	Confidence
Duration of vernalization (Herbaceous)	Increase	Low[a]
Duration of cultivation (Herbaceous, no vernalization)	Decrease	Medium
Duration of cultivation (Herbaceous, obligate vernalization, early varieties)	No change	Low[a]
Duration of cultivation (Herbaceous, obligate vernalization, late varieties)	Decrease	Low[a]
Chilling	Decrease	High
Fruit tree flowering	Advance	High
Blossom frost risk	No increase	High
Irrigation water demand	Increase	Medium
Drought stress	No change	Low
Greenhouse energy consumption	Decrease	High
Heat stress	Increase	Low[b]

[a]Depends largely on the specific vernalization function of the species
[b]Literature review (no projection conducted)

4.4 Critical reflexion

General effects were investigated by detailed projection in case studies. Hence, the presented results are related to the representativeness of the selected systems. For example, only "true" phenophases (Menzel and Sparks 2006: 78) were investigated in detail, whereas the influence of human activity (e.g. technical innovation) nor changes in future variety selection on crop duration were taken into account, hence omitting yield projections. As a consequence, the present work solidifies assumptions about the future boundary conditions for regional horticultural production. Hereby all projections were based on time series, which were compared with or calibrated by observed time series. The latter were in the range of >30 years with two exceptions. First, hourly time series for bias correction development were 10 to 15 years long and found sufficient for method development. Second, blooming time series were in the range of >20 years, as recommended for phenological purposes (Sparks and Menzel 2002). However, climatological means usually refer to periods of ≥ 30 years. Hence, this difference in the latter case must be taken into account.

4.5 Outlook

Climate change impact methodology is changing rapidly, as the manifold members of the simulation chain are renewed continuously. Hence, the presented results can be elaborated with regard to detail and confidence. As seamless seasonal and decadal prediction (Palmer et al. 2008) becomes en vogue, more detailed probabilistic projections can be conducted. These would resemble somewhat probabilistic weather forecasts. Furthermore, Northern Atlantic Oscillation (NAO) patterns are shown to have influences on continental tree phenology (Chmielewski and Rötzer 2001), thus large scale effects could be incorporated in future projections. Finally, the effects of mild winters are not understood to full extent. Hereby mild winters may advance or delay tree phenology, depending on the chilling requirement and photoperiod sensitivity (Körner 2006: 64), and may as well decrease frost hardiness and lead to increased frost risk.

As these notes can only grasp a minor fraction of the investigated area, gaps of knowledge as well as uncertainties in future estimates remain. These can be attributed to plant specific climate responses on the one hand, and climate impact projection methodology on the other hand. Advances in both are required, as both are essential for more detailed statements. Nevertheless "prediction is difficult, especially about the future" (N. Bohr) and thus the presented work can only get a glimpse on the matter at hand. Confidently, future research will lead to a more complete view and eventually be verified by time.

4.5. OUTLOOK

4.3. RÉSUMÉ OF GENERAL EFFECTS

Chapter 5

Bibliography

Abou-Hussein, S. D. (2012). Climate change and its impact on the productivity and quality of vegetable crops (review article). *J Appl Sci Res*, 8(8).

Abu-Asab, M. S., Peterson, P. M., Shetler, S. G., and Orli, S. S. (2001). Earlier plant flowering in spring as a response to global warming in the Washington, DC, area. *Biodivers Conserv*, 10(4):597–612.

Ahuja, I., de Vos, R. C. H., Bones, A. M., and Hall, R. D. (2010). Plant molecular stress responses face climate change. *Trends Plant Sci*, 15(12):664–674.

Amthor, J. S. (2001). Effects of atmospheric CO_2 concentration on wheat yield: Review of results from experiments using various approaches to control CO_2 concentration. *Field Crop Res*, 73(1):1–34.

Baigorria, G. A., Jones, J. W., and O'Brien, J. J. (2008). Potential predictability of crop yield using an ensemble climate forecast by a regional circulation model. *Agric For Meteorol*, 148(8-9):1353–1361.

Bender, S. and Schaller, M. (2012). Vergleichendes Lexikon: Wichtige Definitionen, Schwellenwerte, Kenndaten und Indices für Fragestellungen rund um das Thema "Klimawandel und seine Folgen". Technical report, Climate Service Center.

Berg, P., Feldmann, H., and Panitz, H. . (2012). Bias correction of high resolution regional climate model data. *J Hydrol*, 448-449:80–92.

Bindi, M., Fibbi, L., Gozzini, B., Orlandini, S., and Miglietta, F. (1996). Modeling the impact of future climate scenarios on yield and yield variability of grapevine. *Clim Res*, 7:213–224.

Blanke, M. and Kunz, A. (2009). Einfluss rezenter Klimaveränderungen auf die Phänologie bei Kernobst am Standort Klein-Altendorf - anhand 50-jähriger Aufzeichnungen. *Erwerbs-Obstbau*, 51:101–114.

Blanke, M. M. and Kunz, A. (2011). Effects of climate change on pome fruit phenology and precipitation. *Acta Hortic*, 922:381–386.

Blümel, K. and Chmielewski, F. (2012). Shortcomings of classical phenological forcing models and a way to overcome them. *Agric For Meteorol*, 164:10–19.

Booker, F., Muntifering, R., McGrath, M., Burkey, K., Decoteau, D., Fiscus, E., Manning, W., Krupa, S., Chappelka, A., and Grantz, D. (2009). The ozone component of global change: Potential effects on agricultural and horticultural plant yield, product quality and interactions with invasive species. *J Integr Plant Biol*, 51:337–351.

Bordoy, R. and Burlando, P. (2013). Bias correction of regional climate model simulations in a region of complex orography. *J Appl Meteorol Clim*, 52(1):82–101.

Borochov-Neori, H., Judeinstein, S., Harari, M., Bar-Ya'akov, I., Patil, B. S., Lurie, S., and Holland, D. (2011). Climate effects on anthocyanin accumulation and composition in the pomegranate (Punica granatum L) fruit arils. *J Agr Food Chem*, 59(10):5325–5334.

Botzen, W., Bouwer, L., and van den Bergh, J. (2010). Climate change and hailstorm damage: Empirical evidence and implications for agriculture and insurance. *Resour Energy Econ*, 32:341–362.

Boulard, T., Fatnassi, H., and Tchamitchian, M. (2011). Simulating the consequences of global climate change on greenhouse tomato production in South-France: Preliminary results. *Acta Hortic*, 919:71–80.

Bowes, G. (1991). Growth at elevated CO_2: Photosynthetic responses mediated through Rubisco. *Plant Cell Environ*, 14:795–806.

Brooks, M. M. and Marron, J. S. (1991). Asymptotic optimality of the least-squares cross-validation bandwidth for kernel estimates of intensity functions. *Stoch Proc Appl*, 38(1):157–165.

Campbell, N. and Reece, J. (2003). *Biologie*. Spektrum Akademischer Verlag GmbH, Heidelberg, Berlin.

Campeanu, C., Beleniuc, G., Simionescu, V., Panaitescu, L., and Grigorica, L. (2012). Climate change effects on ripening process and wine composition in Oltenia's vineyards from Romania. *Acta Hortic*, 931:47–54.

Campi, P., Navarro, A., Giglio, L., Palumbo, A., and Mastrorilli, M. (2012). Modelling for water supply of irrigated cropping systems on climate change. *Ital J Agron*, 7(e14):93–99.

Campoy, J. A., Ruiz, D., and Egea, J. (2011). Dormancy in temperate fruit trees in a global warming context: A review. *Sci Hortic-Amsterdam*, 130(2):357–372.

Cannell, G. and Smith, R. (1986). Climatic warming, spring budburst and frost damage on trees. *J Appl Ecol*, 23:177–191.

Chen, C., Haerter, J. O., Hagemann, S., and Piani, C. (2011). On the contribution of statistical bias correction to the uncertainty in the projected hydrological cycle. *Geophys Res Lett*, 38(20).

Chitu, E., Butac, M., and Chitu, V. (2012). Modelling of climatic changes impact on the growth and fruiting of some plum cultivars in the southern part of Romania. *Acta Hortic*, 968:253–260.

CHAPTER 5. BIBLIOGRAPHY

Chmielewski, F., Görgens, M., and Kemfert, C. (2009). KliO: Klimawandel und Obstbau in Deutschland. Abschlussbericht. Technical report, Humboldt University of Berlin, Institute of Crop Sciences, Subdivision of Agricultural Meteorology.

Chmielewski, F., Müller, A., and Bruns, E. (2004). Climate changes and trends in phenology of fruit trees and field crops in Germany, 1961-2000. *Agric For Meteorol*, 121:69–78.

Chmielewski, F., Müller, A., and Küchler, W. (2005). Climate changes and frost hazard for fruit trees. *Annalen der Meteorologie*, 41 2:488–491.

Chmielewski, F. and Rötzer, T. (2001). Response of tree phenology to climate change across Europe. *Agric For Meteorol*, 108:101–112.

Chung, U., Mack, L., Yun, J., and Kim, S.-H. (2011). Predicting the Timing of Cherry Blossoms in Washington, DC and Mid-Atlantic States in Response to Climate Change. *PLOS One*, 6:1–8.

Clarke, A. (2003). Costs and consequences of evolutionary temperature adaptation. *Trends Ecol Evol*, 18(11):573–581.

De Temmerman, L., Wolf, J., Colls, J., Bindi, M., Fangmeier, A., Finnan, J., Ojanperä, K., and Pleijel, H. (2002). Effect of climatic conditions on tuber yield (Solanum tuberosum L.) in the European 'CHIP' experiments. *Eur J Agron*, 17(4):243–255.

Döll, P. (2002). Impact of climate change and variability on irrigation requirements: A global perspective. *Climatic Change*, 54(3):269–293.

Déqué, M., Rowell, D. P., Lüthi, D., Giorgi, F., Christensen, J. H., Rockel, B., Jacob, D., Kjellström, E., De Castro, M., and Van Den Hurk, B. (2007). An intercomparison of regional climate simulations for Europe: Assessing uncertainties in model projections. *Climatic Change*, 81(SUPPL. 1):53–70.

Døving, A. (2009). Climate change and strawberry season in Norway. *Acta Hortic*, 842:753–756.

DWD (2009). Klimawandel im Detail - Zahlen und Fakten zum Klima in Deutschland. DWD-Pressekonferenz am 28. April 2009 in Berlin.

Eccel, E., Rea, R., Caffarra, A., and Crisci, A. (2009). Risk of spring frost to apple production under future climate scenarios: the role of phenological acclimation. *Int J Biometeorol*, 53:273–286.

Ehret, U., Zehe, E., Wulfmeyer, V., Warrach-Sagi, K., and Liebert, J. (2012). Hess opinions "should we apply bias correction to global and regional climate model data?". *Hydrol Earth Syst Sc*, 16(9):3391–3404.

Estrella, N., Sparks, T., and Menzel, A. (2007). Trends and temperature responds in the phenology of crops in Germany. *Glob Change Biol*, 13:1737–1747.

Ewert, F. (2004). Modelling plant responses to elevated CO_2: How important is leaf area index? *Ann Bot-London*, 93(6):619–627.

4.3. RÉSUMÉ OF GENERAL EFFECTS

Ewert, F., Rounsevell, M. D. A., Reginster, I., Metzger, M. J., and Leemans, R. (2005). Future scenarios of European agricultural land use: I. Estimating changes in crop productivity. *Agr Ecosyst Environ*, 107(2-3):101–116.

Fink, M., Kläring, H., and George, E. (2009). Gartenbau und Klimawandel in Deutschland. *Landbauforsch Volk*, 328:1–9.

Fisher, M. (1997). Decline in the juniper woodlands of raydah reserve in southwestern saudi arabia: A response to climate changes? *Global Ecol Biogeogr Lett*, 6(5):379–386.

Førland, E., Benestad, R., Hanssen-Bauer, I., Haugen, J., and Skaugen, T. (2011). Temperature and Precipitation Development at Svalbard 1900-2100. *Advan Meteorol*, pages 1–14.

Friedrich, G. and Fischer, M. (2000). *Physiologische Grundlagen des Obstbaues*. Eugen Ulmer GmbH & Co., Stuttgart (Hohenheim).

Fuhrer, J. (2003). Agroecosystem responses to combinations of elevated CO_2, ozone, and global climate change. *Agr Ecosyst Environ*, 97(1-3):1–20.

Fuhrer, J. (2009). Ozone risk for crops and pastures in present and future climates. *Naturwissenschaften*, 96(2):173–194.

Gebbers, R. (2010). Geostatistics and Kriging (by R. Gebbers). In: Trauth, M. (ed.), *MATLAB® Recipes for Earth Sciences*. Springer, Heidelberg Berlin.

Giorgi, F. and Coppola, E. (2010). Does the model regional bias affect the projected regional climate change? An analysis of global model projections: A letter. *Climatic Change*, 100(3):787–795.

GLOBE Task Team, Hastings, D., Dunbar, P., Elphingstone, G., Bootz, M., Murakami, H., Maruyama, H., Masaharu, H., Holland, P., Payne, J., Bryant, N., Logan, T., Muller, J., Schreier, G., and MacDonald, J. (1999). The Global Land One-kilometer Base Elevation (GLOBE) Digital Elevation Model, Version 1.0. National Oceanic and Atmospheric Administration, National Geophysical Data Center, 325 broadway, Boulder, Colorado 80305-3328, U.S.A., URL = http://www.ngdc.noaa.gov/mgg/topo/globe.html, Last fetch: 20.3.2013.

Gouache, D., Bensadoun, A., Brun, F., Pagé, C., and Makowski, D. (2013). Modelling climate change impact on *Septoria tritici* blotch (STB) in France: Accounting for climate model and disease model uncertainty. *Agric For Meteorol*, 170:242–252.

Gutjahr, O. and Heinemann, G. (2013). Comparing precipitation bias correction methods for high-resolution regional climate simulations using COSMO-CLM - Effects on extreme values and climate change signal. *Theor Appl Climatol*, pages 1–19. Article in Press.

Haberlandt, U., Belli, A., and Hölscher, J. (2010). Trends in beobachteten Zeitreihen von Temperatur und Niederschlag in Niedersachsen (Trends in observed time series of temperature and precipitation in Lower Saxony). *Hydrol Wasserbewirts*, 54:28–36.

CHAPTER 5. BIBLIOGRAPHY

Haerter, J. O., Hagemann, S., Moseley, C., and Piani, C. (2011). Climate model bias correction and the role of timescales. *Hydrol Earth Syst Sc*, 15(3):1065–1079.

Hagemann, S., Chen, C., Haerter, J. O., Heinke, J., Gerten, D., and Piani, C. (2011). Impact of a statistical bias correction on the projected hydrological changes obtained from three GCMs and two hydrology models. *J Hydrometeorol*, 12(4):556–578.

Hantel, M., Kraus, H., and Schönwiese, C.-D. (1987). Climate definition. In: Fischer, G. (ed.), *Landolt-Börnstein Numerical Data and Functional Relationships in Science and Technology, Subvol. c1, Climatology*. Springer Verlag, Berlin.

Harvell, C. D., Mitchell, C. E., Ward, J. R., Altizer, S., Dobson, A. P., Ostfeld, R. S., and Samuel, M. D. (2002). Climate warming and disease risks for terrestrial and marine biota. *Science*, 296(5576):2158–2162.

Hawkins, E., Osborne, T. M., Ho, C. K., and Challinor, A. J. (2013). Calibration and bias correction of climate projections for crop modelling: An idealised case study over Europe. *Agr For Meteorol*, 170:19–31.

Hawkins, E. and Sutton, R. (2009). The potential to narrow uncertainty in regional climate predictions. *B Am Meteorol Soc*, 90:1095–1107.

Hawkins, E. and Sutton, R. (2012). Time of emergence of climate signals. *Geophys res lett*, 39:1–7.

Hebbar, K. B., Venugopalan, M. V., Prakash, A. H., and Aggarwal, P. K. (2013). Simulating the impacts of climate change on cotton production in india. *Climatic Change*, pages 1–13.

Hedhly, A., Hormaza, J. I., and Herrero, M. (2007). Warm temperatures at bloom reduce fruit set in sweet cherry. *J Appl Bot Food Qual*, 81(2):158–164.

Ho, C. K., Stephenson, D. B., Collins, M., Ferro, C. A. T., and Brown, S. J. (2012). Calibration strategies: A source of additional uncertainty in climate change projections. *B Am Meteorol Soc*, 93(1):21–26.

Hoffmann, H., Langner, F., and Rath, T. (2012). Simulating the influence of climatic warming on future spring frost risk in Northern German fruit production. *Acta Hortic*, 957:289–296.

Hoffmann, H. and Rath, T. (2009). Überregionale Simulationen zum zukünftigen Energieverbrauch von Gewächshäusern unter Berücksichtigung von IPCC-Szenarien. *Gesellschaft für Informatik in der Landwirtschaft (GIL), Tagungsband*, 21:61–64.

Hoffmann, H. and Rath, T. (2010). Abiotische Schäden in der Obst- und Gemüseproduktion. Poster. KLIFF-Statusseminar, Göttingen, Germany.

Hoffmann, H. and Rath, T. (2011). Verwendbarkeit simulierter Klimazeitreihen für Pflanzenwachstumsmodelle. *DGG-Proceedings*, 1:1–5.

Hoffmann, H. and Rath, T. (2012a). High resolved simulation of climate change impact on greenhouse energy consumption in Germany. *Eur J Hortic Sci*, 77(6):241–248.

Hoffmann, H. and Rath, T. (2012b). Meteorologically consistent bias correction of simulated climate time series for agricultural models. *Theor Appl Climatol*, 110:129–141.

Hoffmann, H. and Rath, T. (2013a). Abschätzung zukünftiger Klimawandelfolgen im Gartenbau. *BHGL-Schriftenreihe. ISSN 1613-088X*, 29:48.

Hoffmann, H. and Rath, T. (2013b). Estimating the sensitivity and variability of climate impact projections for horticultural models. *European Climate Change Adaptation Conference*. ISBN 978-92-79-26185-5, pages 99–100.

Holmer, B. (2008). Fluctuations of winter wheat yields in relation to length of winter in Sweden 1866 to 2006. *Clim Res*, 36(3):241–252.

Houghton, J., Callender, B., and Varney, S. (1992). The Supplementary Report to the IPCC Scientific Assessment. Technical report, Intergovernmental Panel on Climate Change.

Houghton, J., Ding, Y., Griggs, D., Noguer, M., van der Linden, P., Dai, X., Maskell, K., and Johnson, C. (2001). IPCC Third Assessment Report - Climate Change 2001, Working Group I: The Scientific Basis. Technical report, Intergovernmental Panel on Climate Change.

Hughes, L. (2000). Biological consequences of global warming: Is the signal already apparent? *Trends Ecol Evol*, 15(2):56–61.

Ines, A. V. M. and Hansen, J. W. (2006). Bias correction of daily GCM rainfall for crop simulation studies. *Agr For Meteorol*, 138(1-4):44–53.

Jackson, J. (2003). *Biology of Apples and Pears*. Cambridge University Press, Cambridge.

Jacob, D. (2001). A note to the simulation of the annual and inter-annual variability of the water budget over the baltic sea drainage basin. *Meteorol Atmos Phys*, 77(1-4):61–73.

Jacob, D., Bülow, K., Kotova, L., Moseley, C., Petersen, J., and Rechid, D. (2012). CSC Report 6: Regionale Klimaprojektionen für Europa und Deutschland: Ensemble-Simulationen für die Klimafolgenforschung. Technical report, Climate Service Center, Max Planck Institute for Meteorology.

Janis, M. J. and Robeson, S. M. (2004). Determining the spatial representativenesss of air-temperature records using variogram-nugget time series. *Phys Geogr*, 25(6):513–530.

Juroszek, P. and v. Tiedemann, A. (2011). Potential strategies and future requirements for plant disease management under a changing climate. *Plant Pathol*, 60(1):100–112.

Juroszek, P. and v. Tiedemann, A. (2012). Climate change and potential future risks through wheat diseases: a review. *Eur J Plant Pathol*, 136(1):21–33.

Kampuss, K., Strautina, S., and Laugale, V. (2009). Influence of climate change on berry crop growing in Latvia. *Acta Hortic*, 838:45–50.

CHAPTER 5. BIBLIOGRAPHY

Katul, G. G., Ellsworth, D. S., and Lai, C. . (2000). Modelling assimilation and intercellular CO_2 from measured conductance: A synthesis of approaches. *Plant Cell Environ*, 23(12):1313–1328.

Katz, R. (2002). Techniques for estimating uncertainty in climate change scenarios and impact studies. *Clim Res*, 20(2):167–185.

Kaukoranta, T., Tahvonen, R., and Ylämäki, A. (2010). Climatic potential and risks for apple growing by 2040. *Agr Food Sci*, 19:144–159.

Kerr, R. A. (2013). Forecasting regional climate change flunks its first test. *Science*, 339(6120):638.

Kimball, B. A. and Idso, S. B. (1983). Increasing atmospheric CO_2: effects on crop yield, water use and climate. *Agr Water Manage*, 7(1-3):55–72.

Kling, H., Fuchs, M., and Paulin, M. (2012). Runoff conditions in the upper Danube basin under an ensemble of climate change scenarios. *J Hydrol*, 424-425:264–277.

Köppen, W. (1923). *Die Klimate der Erde*. Walter de Gruyter, Berlin.

Körner, C. (2003). Carbon limitation in trees. *J Ecol*, 91(1):4–17.

Körner, C. (2006). Significance of temperature in Plant Life. In: Morison, JIL and Morecroft, MD (ed.), *Plant Growth and Climate Change*. Blackwell Publishing Ltd, Oxford, UK.

Krug, H., Liebg, H., and Stützel, H., editors (2002). *Gemüseproduktion*. Eugen Ulmer GmbH & Co., Stuttgart (Hohenheim).

Krug, H., Romey, A., and Rath, T. (2007). Decision support for climate dependent greenhouse production planning and climate control by modelling. I. Modelling climate. *Eur J Hortic Sci*, 72(3):97–103.

Kunz, A. and Blanke, M. (2011). Effects of Global Climate Change on Apple 'Golden Delicious' Phenology – Based on 50 Years of Meteorological and Phenological Data in Klein-Altendorf. *Acta Hortic*, 903:1121–1126.

Laapas, M., Jylhä, K., and Tuomenvirta, H. (2012). Climate change and future overwintering conditions of horticultural woody-plants in Finland. *Boreal Environ Res*, 17:31–45.

Latif, M. (2009). *Klimawandel und Klimadynamik*. Eugen Ulmer KG, Stuttgart, Hohenheim.

Leander, R. and Buishand, T. A. (2007). Resampling of regional climate model output for the simulation of extreme river flows. *J Hydrol*, 332(3-4):487–496.

Legave, J., Farrera, I., Almeras, T., and Calleja, M. (2008a). Selecting models of apple flowering time and understanding how global warming has had an impact on this trait. *J Hortic Sci Biotech*, 83:76–84.

Legave, J. M. and Clauzel, G. (2006). Long-term evolution of flowering time in apricot cultivars grown in Southern France: Which future impacts of global warming? *Acta Hortic*, 717:47–50.

4.3. RÉSUMÉ OF GENERAL EFFECTS

Legave, J. M., Regnard, J. ., Farrera, I., Alméras, T., and Calleja, M. (2008b). The modelling of flowering time in the french apple cropping area in relation to global warming. *Acta Hortic*, 772:167-174.

Lenderink, G., Buishand, A., and Van Deursen, W. (2007). Estimates of future discharges of the river rhine using two scenario methodologies: Direct versus delta approach. *Hydrol Earth Syst Sc*, 11(3):1145-1159.

Li, H., Sheffield, J., and Wood, E. F. (2010). Bias correction of monthly precipitation and temperature fields from Intergovernmental Panel on Climate Change AR4 models using equidistant quantile matching. *J Geophys Res Atmos*, 115(10).

Lobell, D. B. and Field, C. B. (2007). Global scale climate-crop yield relationships and the impacts of recent warming. *Environ Res Lett*, 2(1).

Lobell, D. B., Field, C. B., Cahill, K. N., and Bonfils, C. (2006). Impacts of future climate change on California perennial crop yields: Model projections with climate and crop uncertainties. *Agr For Meteorol*, 141(2-4):208-218.

Long, S., Ainsworth, E., Leakey, A., Nösbsrger, J., and Ort, D. (2006). Food for thought: Lower-than-expected crop yield stimulation with rising CO_2 concentrations. *Science*, 312(5782):1918-1921.

Lorencová, E., Frélichová, J., Nelson, E., and Vačkář, D. (2013). Past and future impacts of land use and climate change on agricultural ecosystem services in the Czech Republic. *Land Use Policy*, 33:183-194.

Lorenz, E. (1970). Climatic change as a mathematical problem. *J Appl Meteorol*, 9:325-329.

Lorenz, E. (1975). Climatic predictability. In: WMO-ICSU Joint Organizing Committee: The physical basis of climate and climate modeling. *GARÜ Publ, Appendix 2.1*, 16:132-136.

LSKN (2013). Landesbetrieb für Statistik und Kommunikationstechnologie Niedersachsen. http://www.nls.niedersachsen.de/Tabellen/Landwirtschaft/ernte03/ernte03.htm. Last fetch: 10.04.2013.

Luedeling, E. (2012). Climate change impacts on winter chill for temperate fruit and nut production: A review. *Sci Hortic-Amsterdam*, 144:218-229.

Luedeling, E., Blanke, M., and Gebauer, J. (2009a). Auswirkungen des Klimawandels auf die Verfügbarkeit + Kältewirkungen (Chilling) für Obstgehölze in Deutschland. *Erwerbs-Obstbau*, 51:81-94.

Luedeling, E., Zhang, M., and Girvetz, E. H. (2009b). Climatic changes lead to declining winter chill for fruit and nut trees in California during 1950-2099. *PLOS ONE*, 4(7).

Malheiro, A. C., Santos, J. A., Fraga, H., and Pinto, J. G. (2012). Future scenarios for viticultural climatic zoning in Iberia. *Acta Hortic*, 931:55-62.

Maraun, D., Wetterhall, F., Ireson, A. M., Chandler, R. E., Kendon, E. J., Widmann, M., Brienen, S., Rust, H. W., Sauter, T., Themel, M., Venema, V. K. C., Chun, K. P., Goodess, C. M., Jones, R. G., Onof, C., Vrac,

CHAPTER 5. BIBLIOGRAPHY

M., and Thiele-Eich, I. (2010). Precipitation downscaling under climate change: Recent developments to bridge the gap between dynamical models and the end user. *Rev Geophys*, 48(3).

Marino, S., Hogue, I. B., Ray, C. J., and Kirschner, D. E. (2008). A methodology for performing global uncertainty and sensitivity analysis in systems biology. *J Theor Biol*, 254(1):178–196.

Marris, E. (2007). Gardening: A garden for all climates. *Nature*, 450(7172):937–939.

Marschner, H. (1995). *Mineral Nutrition of Higher Plants*. Academic Press Limited, Cambridge.

Maslin, M. (2013). Cascading uncertainty in climate change models and its implications for policy. *Geogr J*, 179(3):264–271.

Maslin, M. and Austin, P. (2012). Uncertainty: Climate models at their limit? *Nature*, 486(7402):183–184.

Menzel, A. and Sparks, T. (2006). Temperature and plant development: phenology and seasonality. In: Morison, JIL and Morecroft, MD (ed.), *Plant Growth and Climate Change*. Blackwell Publishing Ltd, Oxford, UK.

Miller-Rushing, A., Katsuki, T., Primack, R., Ishii, Y., Lee, S., and Higuchi, H. (2007). Impact of global warming on a group of related species and their hybrids: Cherry tree (*Rosaceae*) flowering at Mt. Takao, Japan. *Am J Bot*, 94:1470–1478.

Moretti, C. L., Mattos, L. M., Calbo, A. G., and Sargent, S. A. (2010). Climate changes and potential impacts on postharvest quality of fruit and vegetable crops: A review. *Food Res Int*, 43(7):1824–1832.

Morison, J. and Morecroft, M. (2006). *Plant Growth and Climate Change*. Blackwell Publishing Ltd, Oxford, UK.

Morison, J. I. L. and Lawlor, D. W. (1999). Interactions between increasing CO_2 concentration and temperature on plant growth. *Plant Cell Environ*, 22(6):659–682.

Moseley, M., Panferov, O., Döring, C., Dietrich, J., Haberlandt, U., Ebermann, V., Rechid, D., Beese, F., and Jacob, D. (2012). Empfehlungen für eine niedersächsische Strategie zur Anpassung an die Folgen des Klimawandels. Niedersächsisches Ministerium für Umwelt, Energie und Klimaschutz, Regierungskommission Klimaschutz.

Moss, C. and De Bodisco, C. (2002). Irrigation projections in Georgia's Alabama-Coosa-Tallapoosa and Apalachicola-Chattahoochee-Flint Basins: 1995-2020. *Water Resour Manag*, 16(5):381–400.

Moss, R. H., Edmonds, J. A., Hibbard, K. A., Manning, M. R., Rose, S. K., Van Vuuren, D. P., Carter, T. R., Emori, S., Kainuma, M., Kram, T., Meehl, G. A., Mitchell, J. F. B., Nakicenovic, N., Riahi, K., Smith, S. J., Stouffer, R. J., Thomson, A. M., Weyant, J. P., and Wilbanks, T. J. (2010). The next generation of scenarios for climate change research and assessment. *Nature*, 463(7282):747–756.

4.3. RÉSUMÉ OF GENERAL EFFECTS

MPI (2006). Hinweise für REMO-Datennutzer. Max Planck Institute for Meteorology. URL: http://www.remo-rcm.de/FAQ-Daten-technisches.1234.0.html (last fetch: 6.4.2013).

Mudelsee, M., Börngen, M., Tetzlaff, G., and Grünewald, U. (2004). Extreme floods in central Europe over the past 500 years: Role of cyclone pathway "Zugstrasse Vb". *J Geophys Res-Atmos*, 109(23):1–21.

Mudelsee, M., Chirila, D., Deutschländer, T., Döring, C., Haerter, J., Hagemann, S., Hoffmann, H., Jacob, D., Krahé, P., Lohmann, G., Moseley, C., Nilson, E., Panferov, O., Rath, T., and Tinz, B. (2010). Climate Model Bias Correction und die Deutsche Anpassungsstrategie. *Mitteilunge DMG*, 3:2–7.

Nakicenovic, N., Alcamo, J., Davis, G., de Vries, B., Fenhann, J., Gaffin, S., Gregory, K., Grübler, A., Jung, T., and Kram, T. (2000). Special Report on Emission Scenarios. Technical report, Intergovernmental Panel on Climate Change.

National Geophysical Data Center (2013). URL = http://www.ngdc.noaa.gov/mgg/shorelines/shorelines.html. Last fetch: 20.3.2013.

Olesen, J., Trnka, M., Kersebaum, K., Skjelvag, A., Seguin, B., Peltonen-Sainio, P., Rossi, F., Kozyra, J., and Micale, F. (2011). Impacts and adaptation of European crop production systems to climate change. *Eur J Agron*, 34:96–112.

Olesen, J. E. and Bindi, M. (2002). Consequences of climate change for European agricultural productivity, land use and policy. *Eur J Agron*, 16(4):239–262.

Olesen, J. E., Børgesen, C. D., Elsgaard, L., Palosuo, T., Rötter, R. P., Skjelvåg, A. O., Peltonen-Sainio, P., Börjesson, T., Trnka, M., Ewert, F., Siebert, S., Brisson, N., Eitzinger, J., van Asselt, E. D., Oberforster, M., and van der Fels-Klerx, H. J. (2012). Changes in time of sowing, flowering and maturity of cereals in Europe under climate change. *Food Addit Contam A*, 29(10):1527–1542.

Olesen, J. E., Carter, T. R., Díaz-Ambrona, C. H., Fronzek, S., Heidmann, T., Hickler, T., Holt, T., Minguez, M. I., Morales, P., Palutikof, J. P., Quemada, M., Ruiz-Ramos, M., Rubæk, G. H., Sau, F., Smith, B., and Sykes, M. T. (2007). Uncertainties in projected impacts of climate change on European agriculture and terrestrial ecosystems based on scenarios from regional climate models. *Climatic Change*, 81(SUPPL. 1):123–143.

Oliver, M. A. and Webster, R. (1990). Kriging: a method of interpolation for geographical information systems. *Int J Geogr Inf Syst*, 4(3):313–332.

Olson, D. M., Dinerstein, E., Wikramanayake, E. D., Burgess, N. D., Powell, G. V. N., Underwood, E. C., D'Amico, J. A., Itoua, I., Strand, H. E., Morrison, J. C., Loucks, C. J., Allnutt, T. F., Ricketts, T. H., Kura, Y., Lamoreux, J. F., Wettengel, W. W., Hedao, P., and Kassem, K. R. (2001). Terrestrial ecoregions of the world: A new map of life on Earth. *Bioscience*, 51(11):933–938.

Osborne, T., Rose, G., and Wheeler, T. (2013). Variation in the global-scale impacts of climate change on crop productivity due to climate model uncertainty and adaptation. *Agr Forest Meteorol*, 170:183–194.

CHAPTER 5. BIBLIOGRAPHY

Palmer, T. N., Doblas-Reyes, F. J., Weisheimer, A., and Rodwell, M. J. (2008). Toward seamless prediction: Calibration of climate change projections using seasonal forecasts. *B Am Meteorol Soc*, 89(4):459–470.

Parry, M., Canziani, O., Palutikof, J., van der Linden, P., and Hanson, C. (2007). Contribution of Working Group II to the Fourth Assessment Report of the Intergovernmental Panel on Climate Change. Technical report, Intergovernmental Panel on Climate Change.

Piani, C. and Haerter, J. (2012). Two dimensional bias correction of temperature and precipitation copulas in climate models. *Geophys Res Lett*, 39.

Piani, C., Haerter, J., and Coppola, E. (2010). Statistical bias correction for daily precipitation in regional climate models over Europe. *Theor Appl Climatol*, 99:187–192.

Pielke, R., Stohlgren, T., Parton, W., Doesken, N., Moeny, J., Schell, L., and Redmond, K. (2000). Spatial representativeness of temperature measurements from a single site. *B Am Meteorol Soc*, 81(4):826–830.

Pieri, P., Lebon, E., and Brisson, N. (2012). Climate change impact on french vineyards as predicted by models. *Acta Hortic*, 931:29–38.

PIK (2013). Potsdam Institute for Climate Impact Research, URL=http://www.pik-potsdam.de/services/climate-weather-potsdam/climate-diagrams/global-radiation. Last fetch: 18.3.2013.

Porter, J. R. and Semenov, M. A. (2005). Crop responses to climatic variation. *Philos T Roy Soc B*, 360(1463):2021–2035.

Quebedeaux, B. and Chollet, R. (1977). Comparative growth analyses of *Panicum* species with differing rates of photorespiration. *Plant Physiol*, 59:42–44.

Raper, C. and Kramer, P., editors (1983). *Crop Reactions to Water and Temperature Stresses in Humid, Temperate Climates*. Westview Press, Boulder, Colorado, United States.

Rasmussen, J., Sonnenborg, T. O., Stisen, S., Seaby, L. P., Christensen, B. S. B., and Hinsby, K. (2012). Climate change effects on irrigation demands and minimum stream discharge: Impact of bias-correction method. *Hydrol Earth Syst Sc*, 16(12):4675–4691.

Rehana, S. and Mujumdar, P. P. (2012). Regional impacts of climate change on irrigation water demands. *Hydrol Process*. Article in Press.

Reichler, T. and Kim, J. (2008). How well do coupled models simulate today's climate? *B Am Meteorol Soc*, 89(3):303–311.

Reidsma, P. and Ewert, F. (2008). Regional farm diversity can reduce vulnerability of food production to climate change. *Ecol Soc*, 13(1).

Richter, D. (1995). *Ergebnisse methodischer Untersuchungen zur Korrektur des systematischen Meßfehlers des Hellmann-Niederschlagsmessers. Berichte des Deutschen Wetterdienstes 194*. Deutscher Wetterdienst, Offenbach am Main.

4.3. RÉSUMÉ OF GENERAL EFFECTS

Räisänen, J. and Ruokolainen, L. (2008). Estimating present climate in a warming world: A model-based approach. *Climate Dynamics*, 31(5):573–585.

Rochette, P., Bélanger, G., Castonguay, Y., Bootsma, A., and Mongrain, D. (2004). Climate change and winter damage to fruit trees in eastern Canada. *Can J Plant Sci*, 84(4):1113–1125.

Rodrigo, J. (2000). Spring frosts in deciduous fruit trees – morphological damage and flower hardiness. *Sci Hortic-Amsterdam*, 85:155–173.

Roeckner, E., Bäuml, G., Bonaventura, L., Brokopf, R., Esch, M., Giorgetta, M., Hagemann, S., Kirchner, I., Kornblueh, L., Manzini, E., Rhodin, A., Schlese, U., Schulzweida, U., and Tompkins, A. (2003). The atmospheric general circulation model ECHAM5, Part 1, Model description. Technical report, Max Planck Institute for Meteorology.

Rolland, C. (2003). Spatial and seasonal variations of air temperature lapse rates in alpine regions. *J Climate*, 16(7):1032–1046.

Rosenzweig, C., Phillips, J., Goldberg, R., Carroll, J., and Hodges, T. (1996). Potential Impacts of Climate Change on Citrus and Potato Production in the US. *Agr Syst*, 52:455–479.

Rosenzweig, C., Tubiello, F. N., Goldberg, R., Mills, E., and Bloomfield, J. (2002). Increased crop damage in the US from excess precipitation under climate change. *Global Environ Chang*, 12(3):197–202.

Rötter, R. and Van De Geijn, S. C. (1999). Climate change effects on plant growth, crop yield and livestock. *Climatic Change*, 43(4):651–681.

Rötter, R. P., Palosuo, T., Pirttioja, N. K., Dubrovsky, M., Salo, T., Fronzek, S., Aikasalo, R., Trnka, M., Ristolainen, A., and Carter, T. R. (2011). What would happen to barley production in Finland if global warming exceeded 4 °C? A model-based assessment. *Eur J Agron*, 35(4):205–214.

Rubino, P., Stelluti, M., Stellacci, A., Armenise, E., Ciccarese, A., and Sellami, M. (2012). Yield response and optimal allocation of irrigation water under actual and simulated climate change scenarios in a southern Italy district. *Ital J Agron*, 7(e18):124–132.

Rugienius, R., Siksnianas, T., Gelvonauskiene, D., Staniene, G., Sasnauskas, A., Zalunskaite, I., and Stanys, V. (2009). Evaluation of genetic resources of fruit crops as donors of cold and disease resistance in Lithuania. *Acta Hortic*, 825:117–124.

Ruosteenoja, K., Tuomenvirta, H., and Jylhä, K. (2007). GCM-based regional temperature and precipitation change estimates for Europe under four SRES scenarios applying a super-ensemble pattern-scaling method. *Climatic Change*, 81(SUPPL. 1):193–208.

Sage, R. F., Way, D. A., and Kubien, D. S. (2008). Rubisco, rubisco activase, and global climate change. *J Exp Bot*, 59(7):1581–1595.

CHAPTER 5. BIBLIOGRAPHY

Santos, J. A., Grätsch, S. D., Karremann, M. K., Jones, G. V., and Pinto, J. G. (2013). Ensemble projections for wine production in the Douro Valley of Portugal. *Climatic Change*, 117(1-2):211–225.

Santos, J. A., Leite, M. S., Bennett, R. N., and Rosa, E. A. S. (2012). The changing climate: Using modeling to predict potential effects on horticultural crops. *Acta Hortic*, 936:89–94.

Sato, S., Kamiyama, M., Iwata, T., Makita, N., Furukawa, H., and Ikeda, H. (2006). Moderate increase of mean daily temperature adversely affects fruit set of lycopersicon esculentum by disrupting specific physiological processes in male reproductive development. *Ann Bot-London*, 97(5):731–738.

Schaller, M. and Weigel, H. (2007). Analyse des Sachstands zu Auswirkungen von Klimaveränderungen auf die deutsche Landwirtschaft und Maßnahmen zur Anpassung. *Landbauforsch Volk*, 316:1–252.

Scheffer, F. and Schachtschabel, P. (1989). *Bodenkunde*. Ferdinand Enke Verlag, Stuttgart.

Scheifinger, H., Menzel, A., Koch, E., and Peter, C. (2003). Trends in spring time frost events and phenological dates in Central Europe. *Theor Appl Climatol*, 74:41–51.

Schmidli, J., Frei, C., and Vidale, P. L. (2006). Downscaling from GCM precipitation: A benchmark for dynamical and statistical downscaling methods. *Int J Climatol*, 26(5):679–689.

Schonwiese, C. and Janoschitz, R. (2008). *Klima-Trendatlas Deutschland 1901-2000. 2. aktualisierte Auflage, Bericht Nr.4*. Inst. Atm. Umwelt, Univ. Frankfurt, Frankfurt.

Schroeter, D., Cramer, W., Leemans, I., Prentice, I., Araújo, M., Arnell, N., Bondeau, A., Bugmann, H., Carter, T., Gracia, C., de la Vega-Leinert, A., Erhard, M., Ewert, F., Glendining, M., House, J., Krankaanpää, S., Klein, R., Lvaorel, S., Lindner, M., Metzger, M., Meyer, J., Mitchell, T., Reginster, I., Rounsevell, M., Sabaté, S., Sitch, S., Smith, B., Smith, J., Smith, P., Sykes, M., Thonicke, K., Thuiller, W., Tuck, G., Zaehle, S., and Zierl, B. (2005). Ecosystem Service Supply and Vulnerability to Global Change in Europe (Supporting online Material). *Science*, 310:1333–1337.

Schultz, H. R. (2000). Climate change and viticulture: A European perspective on climatology, carbon dioxide and UV-B effects. *Aust J Grape Wine R*, 6(1):2–12.

Seaby, L. P., Refsgaard, J. C., Sonnenborg, T. O., Stisen, S., Christensen, J. H., and Jensen, K. H. (2013). Assessment of robustness and significance of climate change signals for an ensemble of distribution-based scaled climate projections. *J Hydrol*. Article in Press.

Semenov, M. A. and Shewry, P. R. (2011). Modelling predicts that heat stress, not drought, will increase vulnerability of wheat in Europe. *Scientific Reports*, 1.

Simpson, G. (1981). *Water Stress on Plants*. Praeger Publishers, New York.

Solomon, S., Quin, D., Manning, M., Chen, Z., Marquis, M., Averyt, K., Tignor, M., and Miller, H. (2007). The Physical Science Basis. Contribution of Working Group I to the Fourth Assessment Report of the Intergovernmental Panel on Climate Change. Technical report, Intergovernmental Panel on Climate Change.

Soora, N. K., Aggarwal, P. K., Saxena, R., Rani, S., Jain, S., and Chauhan, N. (2013). An assessment of regional vulnerability of rice to climate change in India. *Climatic Change*, pages 1–17.

Sparks, T. H. and Menzel, A. (2002). Observed changes in seasons: An overview. *Int J Climatol*, 22(14):1715–1725.

Stöckle, C. O., Marsal, J., and Villar, J. M. (2011). Impact of climate change on irrigated tree fruit production. *Acta Hortic*, 889:41–52.

Sugiura, T. (2010). Characteristics of responses of fruit trees to climate changes in Japan. *Acta Hortic*, 872:85–88.

Taiz, L. and Zeiger, E. (2000). *Physiologie der Pflanzen. Original title: Plant Physiology.* Spektrum Akademischer Verlag GmbH, Heidelberg, Berlin.

Tans, P. (2013). National Geophysical Data Center, URL = http://www.esrl.noaa.gov/gmd/ccgg/trends/. Last fetch: 20.3.2013.

Tebaldi, C. and Lobell, D. (2008). Towards probalistic projections of climate change impacts on global crop yield. doi=10.1029/2008gl033423. *Geophys Res Lett*, 35.

Teixeira, E. I., Fischer, G., Van Velthuizen, H., Walter, C., and Ewert, F. (2013). Global hot-spots of heat stress on agricultural crops due to climate change. *Agr Forest Meteorol*, 170:206–215.

Teutschbein, C. and Seibert, J. (2012). Bias correction of regional climate model simulations for hydrological climate-change impact studies: Review and evaluation of different methods. *J Hydrol*, 456-457.

Trnka, M., Olesen, J. E., Kersebaum, K. C., Skjelvåg, A. O., Eitzinger, J., Seguin, B., Peltonen-Sainio, P., Rötter, R., Iglesias, A., Orlandini, S., Dubrovský, M., Hlavinka, P., Balek, J., Eckersten, H., Cloppet, E., Calanca, P., Gobin, A., Vučetić, V., Nejedlik, P., Kumar, S., Lalic, B., Mestre, A., Rossi, F., Kozyra, J., Alexandrov, V., Semerádová, D., and Žalud, Z. (2011). Agroclimatic conditions in Europe under climate change. *Glob Change Biol*, 17(7):2298–2318.

Turner, N. and Kramer, P., editors (1980). *Adaptation of Plants to Water and High Temperature Stress.* John Wiley & Sons, Inc., New York.

UBA/MPI (2006). Künftige Klimaänderungen in Deutschland — Regionale Projektionen für das 21. Jahrhundert. Hintergrundpapier April 2006. Max Planck Institute for Meteorology, Umweltbundesamt. URL: http://www.umweltbundesamt.de/uba-info-medien/3552.html. Last fetch: 25.04.2013.

Ulukan, H. (2009). Environmental management of field crops: A case study of Turkish agriculture. *International Journal of Agriculture and Biology*, 11(4):483–494.

USDA (1995). ACT/ACF river basins comprehensive study: Agricultural water demand. U.S. Department of Agriculture/Natural Resources Conservation Service, Soil Conservation Service, Washington DC.

CHAPTER 5. BIBLIOGRAPHY

v. Hann, J. (1883). *Handbuch der Klimatologie*. Engelhorn, Stuttgart.

v. Humboldt, A. (1845). *Kosmos. Entwurf einer physischen Weltbeschreibung, Vol. 1*. Cotta'sche Buchhandlung, Stuttgart.

Van Den Bergh, I., Ramirez, J., Staver, C., W. Turner, D., Jarvis, A., and Brown, D. (2012). Climate change in the subtropics: The impacts of projected averages and variability on banana productivity. *Acta Hortic*, 928:89–100.

van Oldenborgh, G., Doblas Reyes, F., Drijfhout, S., and Hawkins, E. (2013). Reliability of regional climate model trends. *Environ Res Lett*, 8.

van Roosmalen, L., Christensen, J. H., Butts, M. B., Jensen, K. H., and Refsgaard, J. C. (2010). An intercomparison of regional climate model data for hydrological impact studies in Denmark. *J Hydrol*, 380(3-4):406–419.

Villani, G., Tomei, F., Tomozeiu, R., and Marletto, V. (2011). Climatic scenarios and their impacts on irrigated agriculture in Emilia-Romagna, Italy. *Ital J Agrometeorol*, 1:5–16.

Vršic, S. and Vodovnik, T. (2012). Reactions of grape varieties to climate changes in North East Slovenia. *Plant Soil Environ*, 58(1).

Vujadinović, M., Vuković, A., Djurdjević, V., Ranković-Vasić, Z., Atanacković, Z., Sivčev, B., Marković, N., and Petrović, N. (2012). Impact of climate change on growing season and dormant period characteristics for the balkan region. *Acta Hortic*, 931:87–94.

Wand, S. (2007). Vulnerability and impact of climate change on horticultural crop production in the Western Cape Province, South Africa. *S Afr J Bot*, 73:321.

Wand, S., Steyn, W., and Theron, K. (2008). Vulnerability and impact of climate change on pear production in South Africa. *Acta Hortic*, 800 PART 1:263–271.

Watanabe, S., Kanae, S., Seto, S., Yeh, P. J. ., Hirabayashi, Y., and Oki, T. (2012). Intercomparison of bias-correction methods for monthly temperature and precipitation simulated by multiple climate models. *J Geophys Res-Atmos*, 117(23).

Wilby, R. L. and Wigley, T. M. L. (1997). Downscaling general circulation model output: A review of methods and limitations. *Prog Phys Geog*, 21(4):530–548.

Woiwod, I. P. (1997). Detecting the effects of climate change on Lepidoptera. *J Insect Conserv*, 1(3):149–158.

Wolfe, D., Schwartz, M., Lakso, A., Otsuki, Y., Pool, R., and Shaulis, N. (2005). Climate change and shifts in spring phenology of three horticultural woody perennials in northeastern USA. *Int J Biometeorol*, 49:303–309.

4.3. RÉSUMÉ OF GENERAL EFFECTS

Wollenweber, B., Porter, J. R., and Schellberg, J. (2003). Lack of interaction between extreme high-temperature events at vegetative and reproductive growth stages in wheat. *J Agron Crop Sci*, 189(3):142–150.

Wurr, D., Fellows, J., and Fuller, M. (2004). Simulated effects of climate change on the production pattern of winter cauliflower in the UK. *Sci Hortic-Amsterdam*, 101(4).

Wurr, D., Fellows, J., and Phelps, K. (1996). Investigating trends in vegetable crop response to increasing temperature associated with climate change. *Sci Hortic-Amsterdam*, 66(3-4):255–263.

Wurr, D., Hand, D., Edmondson, R., Fellows, J., Hannah, M., and Cribb, D. (1998). Climate change: A response surface study of the effects of CO_2 and temperature on the growth of beetroot, carrots and onions. *J Agr Sci*, 131(2):125–133.

Yip, S., Ferro, C., Stephenson, D., and Hawkins, E. (2011). A simple, coherent framework for partitioning uncertainty in climate predictions. *J Climate*, 24(17):4634–4643.

You, J., Hubbard, K. G., and Goddard, S. (2008). Comparison of methods for spatially estimating station temperatures in a quality control system. *In J Climatol*, 28(6):777–787.

Zebisch, M., Grothmann, T., Schröter, D., Hasse, C., Fritsch, U., and Cramer, W. (2005). Klimawandel in Deutschland: Vulnerabilität und Anpassungsstrategien klimasensitiver Systeme. Umweltforschungsplan des Bundesministeriums für Umwelt, Naturschutz und Reaktorsicherheit. Forschungsbericht 201 41 253, UBA-FB 000844. URL: http://www.umweltdaten.de/publikationen/fpdf-1/2947, last fetch: 22.04.2013.

Zhang, X. and Cai, X. (2011). Climate change impacts on global agricultural land availability. *Environ Res Lett*, 6(1).

Chapter 6

Appendix

6.1 Publications

6.1.1 Publications included in the thesis

Hoffmann H, Rath T, 2013. Future bloom and blossom frost risk for *Malus domestica* considering climate model and impact model uncertainties. PLoS ONE 8 (10): e75033. doi:10.1371/journal.pone.0075033.

Hoffmann H, Rath T, 2012. Meteorologically consistent bias correction of climate time series for agricultural models. Theor Appl Climatol 110, 129-141.

Hoffmann H, Rath T, 2012. High resolved simulation of climate change impact on greenhouse energy consumption in Germany. Eur J Hortic Sci 77, 241-248.

Duncker C, Fricke A, Hoffmann H, Rath T, 2012. Dynamic Modelling of Water Stress for *Lactuca sativa* L. var. *capitata*. Acta Hortic, accepted (9.7.2013).

6.1.2 Publications not included in the thesis and conference contributions

The following publications are related to the thesis and partially cited in the thesis (other publications are not listed).

Proceedings / Short Commun.
Hoffmann H, Langner F, Rath T, 2012. Simulating the influence of climatic warming on future spring frost risk in northern German fruit production. Acta Hortic 957, 289-296.

Hoffmann H, Rath T, 2011. Verwendbarkeit simulierter Klimazeitreihen für Pflanzenwachstumsmodelle. DGG-Proceedings, Vol. 1, No. 2, p. 1-5. DOI: 10.5288/dgg-pr-01-02-hh-2011.

Hoffmann H, Rath T, 2009. Überregionale Simulationen zum zukünftigen Energieverbrauch von Gewächshäusern unter Berücksichtigung von IPCC-Szenarien. GIL 21, 61-64. ISBN 978-3-88579-236-9.

Other
Mudelsee M, Chirila D, Deutschländer T, Döring C, Haerter J, Hagemann S, Hoffmann H, Jacob D, Krahé P, Lohmann G, Moseley C, Nilson E, Panferov O, Rath T, Tinz B, 2010. Climate Model Bias Correction und die Deutsche Anpassungsstrategie. Mitteilungen DMG: 03 / 2010. ISSN 0177-8501.

Conference Presentations
Hoffmann, H., Rath, T., 2013. Abschätzung zukünftiger Klimawandelfolgen im Gartenbau. 49. Jahrestagung DGG, Bonn, Germany, 28.02.2013. ISSN 1613-088X, BHGL-Schriftenreihe 29, 48.

CHAPTER 6. APPENDIX
6.1. PUBLICATIONS

Hoffmann H, Langner F, Rath T, 2012. Simulating the influence of climatic warming on future spring frost in northern German fruit production. Hortimodel, Nanjing, China. Acta Hortic 957, 289-296.

Hoffmann H, Rath T, 2011. Verwendbarkeit simulierter Klimazeitreihen für Pflanzenwachstumsmodelle. 47th DGG, Hannover, Germany. BHGL-Schriftenreihe 28, 32.

Hoffmann H, Rath T, 2009. Überregionale Simulationen zum zukünftigen Energieverbrauch von Gewächshäusern unter Berücksichtigung von IPCC-Szenarien. 29th GIL, Rostock, Germany. GIL 21, 61-64. ISBN 978-3-88579-236-9.

Hoffmann H, Rath T, 2009. Zukünftiger Energieverbrauch bundesdeutscher Gewächshäuser unter Berücksichtigung unterschiedlicher Klimaszenarien. 45th DGG, Berlin, Germany. BHGL-Schriftenreihe 26, 60.

Poster (selection)

Hoffmann, H., Rath, T., 2013. Estimating the sensitivity and variability of climate impact projections for horticultural models. European Climate Change Adaptation Conference, Hamburg, Germany. ISBN 978-92-79-26185-5, doi 10.2777/13121, p. 99-100.

Hoffmann H, Rath T, 2012. Regional climate change impact on future energy consumption of greenhouses. Int. Conference on Agricultural Engineering (CIGR), Valencia, Spain.

Hoffmann H, Rath T, 2011. New meteorologically consistent bias correction of simulated climate time series for multidimensional plant models. UNEP Int. Student Conference on Environment and Sustainability, Shanghai, China.

Hoffmann, H., Rath, T., 2010. Abiotische Schäden in der Obst- und Gemüseproduktion. KLIFF-Statusseminar, Göttingen, Germany.

6.2 Acknowledgments

I gratefully and sincerely thank Prof. Dr. Thomas Rath for his guidance, understanding, patience, and most importantly, his friendship during my studies at Biosystems Engineering, Leibniz Universität Hannover. His unorthodox and yet always upright manner encouraged me to not only grow as a researcher but also as an independent thinker. The opportunity I was given to work with such independence contributed significantly to this work.

I am very grateful to the doctoral committee and wish to thank Prof. Dr. Stützel and Prof. Dr. Hau for their support and engagement.

The cooperation with the Vegetable Systems Modelling team has been fruitful and I am deeply grateful to Dr. Andreas Fricke and Karsten Zutz for their engagement in coordinating the technical issues regarding the drought stress experiments. In this matter the dedication of Charlotte Duncker deserves respect, who persisted in setting up new trials. Thanks Charlotte for being a great student.

I thank all KLIFF project partners for their collaboration. The network cooperation across varying types of institutions has been inspiring and I thank Dr. Stella Aspelmeier, Dr. Peter Juroszek and Prof. Dr. von Tiedemann of the G.-A.-Universität Göttingen for synchronizing all these projects. Additionally, I am grateful to Prof. Dr. Daniela Jacob and Dr. Christopher Moseley of the Climate Service Center / Max Planck Institute for Meteorology for providing data and not backing up on all those questions, especially on how to pass >20 TB of climate data quickly from A to B.

I thank all members of the Biosystems Engineering research group for their invaluable support, infinite patience and in vivo friendship. With Gökhan Akyazi, Dr. Stefanie Grade, Klaus Knösel, Christian Marx, Dr. Sebastian Menke and Dr. Cerebro J. Pastrana I have spent not only the past 3 years, but endless hours of discussion and airily sailed some stormy seas. I thank Dr. von Elsner and Prof. Dr. Tantau for sharing their expertise at all moments. For their support in all technical matters I thank Andreas Meyer, Werner Hock, Norbert Grisat and Niko Gilantzis. For backing me up on cumbersome administration I thank Sigrid Cohrs-Zingel, Edda Thülig and Rita Barth. I wish Anne, Erick, Felix, Frederik, Jones, Sandra and Serge all the best for their future. It was a great experience and a real pleasure to be in such a dynamic and open minded team.

Further, I would like to thank Dr. Paul Cochrane and Dr. Andreas Gerdes of the RRZN cluster team for their assistance on remote clustered unix OS with compiled MatLab.

Finally and most importantly, I thank my wife Rike. Her support, encouragement, patience and unwavering love were the bedrock upon which the past years of my life have been built. Needless to mention, that the present work would not have been realized without her. For supporting me at all times, I thank my parents Rainer and Elfriede. Thanks for all your faith in me.

i want morebooks!

Buy your books fast and straightforward online - at one of world's fastest growing online book stores! Environmentally sound due to Print-on-Demand technologies.

Buy your books online at
www.get-morebooks.com

Kaufen Sie Ihre Bücher schnell und unkompliziert online – auf einer der am schnellsten wachsenden Buchhandelsplattformen weltweit! Dank Print-On-Demand umwelt- und ressourcenschonend produziert.

Bücher schneller online kaufen
www.morebooks.de

 VDM Verlagsservicegesellschaft mbH
Heinrich-Böcking-Str. 6-8　　　Telefon: +49 681 3720 174　　　info@vdm-vsg.de
D - 66121 Saarbrücken　　　　Telefax: +49 681 3720 1749　　　www.vdm-vsg.de

Printed by Books on Demand GmbH, Norderstedt / Germany